Carlos Cordova is Professor of Geography, Oklahoma State University. He has spent almost two decades researching the environmental history of the Crimean Peninsula, including its geoarchaeological, paleoecological and biogeographical aspects. He has also undertaken fieldwork in the Middle East, the USA, Mexico and South Africa and is the author of *Millennial Landscape Change in Jordan: Geoarchaeology and Cultural Ecology* (2007).

ENVIRONMENTAL HISTORY AND GLOBAL CHANGE SERIES

This important new series provides a much needed forum for understanding just how and why our environment changes. It shows how environmental history – with its unique blend of geography, history, archaeology, landscape, environment and science – is helping to make informed decisions on pressing environmental concerns and providing crucial insights into the mechanisms that influence environmental change today. The focus of the series will be on contemporary problems but will also include work that addresses major techniques, key periods and important regions. At a time when the scale and importance of environmental change has led to a widespread feeling that we have entered a period of crisis, the *Environmental History and Global Change Series* provides a timely, informed and important contribution to a key global issue.

1. *Documentary Records of Climate Change*, Astrid Ogilvie
2. *A Dictionary of Environmental History*, Ian Whyte
3. *The Mediterranean World: An Environmental History*, Neil Roberts
4. *Seeds of Empire: The Environmental Transformation of New Zealand*, Tom Brooking and Eric Pawson
5. *Cities: An Environmental History*, Ian Douglas
6. *Japan: An Environmental History*, Conrad Totman
7. *Crimea and the Black Sea: An Environmental History*, Carlos Cordova

CRIMEA AND THE BLACK SEA

An Environmental History

CARLOS CORDOVA

I.B.TAURIS
LONDON · NEW YORK

Published in 2016 by
I.B.Tauris & Co. Ltd
London • New York
www.ibtauris.com

Environmental History and Global Change: 7

ISBN: 978 1 78453 001 3
eISBN: 978 0 85773 903 2

A full CIP record for this book is available from the British Library
A full CIP record is available from the Library of Congress

Library of Congress Catalog Card Number: available

Typeset in Garamond Three by OKS Prepress Services, Chennai, India

CONTENTS

LIST OF ILLUSTRATIONS

FIGURES

TABLES

ACKNOWLEDGEMENTS

The process of researching Crimea has produced one of my best friends, a former member of the research corpus of the Nikitsky Botanical Gardens, Alexander Rybak, who now lives in the United States. He was a student of the famous landscape ecologist, the late Professor Gryegoriy Y. Grishankov, who fervently studied the habitats and special territories of Crimea. Inspiration for this book came from works on Crimea's natural history by local geographers (such as Podgorodyetskiy 1988; Yena, Yena and Yena 2004b, 2007). Professor Pyotr D. Podgorodyetskiy even encouraged me to study more of Crimea's natural history in a dedication he scribbled on a copy of his book *Crimea's Nature*. Professor Andrey V. Yena, a botanist of the Crimean Agrotechnological University, with whom I have corresponded and exchanged ideas, has been an incredibly helpful colleague in the comments and suggestions he has offered me in the writing of this book. Andrey and his father and brother, Vladimir Yena and Aleksey Yena, have published a number of articles and books about Crimea's natural history, most of which have been a great source of inspiration. The invaluable work by the Yenas on the history of Crimean geography and environmental science (Yena, Yena and Yena 2007) is the most complete of its kind, which is one reason I rely heavily on it, particularly in Chapter 2.

I am first of all indebted to my colleague Professor Volodymyr Pashchenko (a geographer of the National Ukrainian Academy of Science) for introducing me to the geography of the Crimean Peninsula and for his encouragement to me to become a *krymovyed*. Although far from being that, I developed a strong attachment to the Crimean landscape. I am very much indebted to Yekaterina A. Pozachyenyuk of the Faculty of Geography at the Vernadskiy Tavrida National University, with whom I have exchanged ideas and whose studies have been a great support. Likewise, the late geomorphologist Professor Aleksandr A. Klyukin, from the same university, spent several days with me in the field. Part of our joint work was published recently (Cordova et al.

2011). I am also indebted to Lilya V. Bondaryeva, a former graduate of the Nikitsky Botanical Gardens and now staff member at the A.O. Kovalevskiy Institute of Biology of the Southern Seas, Sevastopol, for her help with identification of flora and for numerous e-mail exchanges on matters of botany and nature conservation. Special thanks also to one of my collaborators and colleagues, palynologist Dr Natalia P. Gerasimenko, now at the Taras Shevchenko National University in Kiev, the other half of Crimean palynology. While I struggled with Holocene pollen, she battled with Late Pleistocene pollen, a collaboration that resulted in a summarized history of the late Quaternary vegetation history of Crimea (Cordova et al. 2011). Many thanks also to historian Aleksey V. Ivanov of the museum at Chersonesos, a scholar who contributed so much to my knowledge of Crimean history. My debt in this endeavor also goes to Galina Nikolaenko, her son and daughter Mikhail and Vera, and to Leonid Marchenko, former director of the National Chersonesos Reserve and Museum, all of whom hosted me and introduced me to the archaeological and natural riches of Crimea in the late 1990s.

My special thanks go also to my editor at I.B.Tauris, David Stonestreet, with whom I have had the great pleasure to work, as well as to the anonymous reviewers who helped improve the content of this book. Finally, the writing of this book is not only my act of gratitude to Crimea, its people and to my Crimean colleagues and friends: it is also a way of bringing to wider attention the still poorly known environmental history and issues of Crimea to the academic world.

PREFACE

In North America, the name Crimea is often associated with the Crimean War (1853–4). Some Americans may have read the war reports in Leo Tolstoy's *Sevastopol Sketches*, or Mark Twain's *Innocents Abroad*, where he described a brief visit to Sevastopol. Some may be familiar with the 248-day siege of Sevastopol during World War II and the 1945 Yalta Conference, where the leaders of the victorious Allied Superpowers—Winston Churchill, Franklin D. Roosevelt and Joseph Stalin—convened to discuss the future of postwar Europe. But many would not know that Sevastopol and Yalta are two Crimean cities. Many educated people in North America may not even know where Crimea is, or what country it is part of, let alone know about its natural and cultural riches. In western Europe, although there is more awareness about what lies to the east, little is known about the geography and the natural and cultural attractions of Crimea, or even what this peninsula in the Black Sea has to offer in terms of tourism. Under these circumstances, the news about the annexation of Crimea by Russia in March 2014 took many by surprise. In reality, Crimea had not been in the world news since 1945.

On the other hand, for citizens of Russia, Ukraine and other former Soviet republics, Crimea represents the top summer vacation destination and is a place of abundant sun, beautiful mountain landscapes, and high-quality wines, champagne and cognac. Rarely, however, do tourists from other countries make Crimea their vacation destination. Tourist guidebooks and websites about Crimea, its cities, parks, trails, museums and botanical gardens abound, but they are predominantly in Russian and Ukrainian. Likewise, the number of scientific books on Crimea's geography, geology, flora and fauna, ecosystems, archaeology and history could fill entire rooms, but again they are mainly in Russian and Ukrainian. Knowledge of Crimean natural history, cultural history, ethnography and geography is so vast that an informal specialization on Crimea exists. Thus names such as *krymovyedyeniye* and *krymovyed*, roughly translated from Russian as Crimean studies and

Crimean specialist, respectively, reflect the special character of the peninsula as a region and as a scholarly discipline in its own right. Similarly, names such as *krymustroystvo* (Crimean system) and *krymologiya* (Crimealogy) appear in websites and social networking sites.

The number of publications (articles, books and chapters in edited volumes) on Crimea's natural history, geography, environmental history and conservation is vast. However, language barriers and the scarcity of these books in libraries outside the former Soviet republics have presented a problem for scholars in the West who are interested in the peninsula. Even more difficult to access are the numerous reports, theses and documents with limited circulation, which are available only in local libraries and institutions, although in recent years the internet has eased these difficulties a little.

How many books on Crimea's geography, natural and environmental history exist in languages other than Russian and Ukrainian? Many books were published in German, French, English and Italian by eighteenth- and nineteenth-century naturalists and travelers, some of whom are cited in Chapter 2. However, only two books on the natural history and geography of Crimea were published in western European languages in the twentieth century. One, in German, was Heinrich Walter's 1943 *Die Krim*, which was meant to be a report of his research on the natural resources of Crimea during the Nazi occupation of the peninsula. The other one, in English, was the Biodiversity Support Program's 1999 volume titled *Priority-Setting in Conservation: A New Approach for Crimea*, which is a technical report focusing on the status of endangered species and conservation territories of the peninsula.

It is difficult to explain why very few books on Crimea's natural and cultural history are written in languages other than Russian and Ukrainian. One reason may be the isolation of Crimea from foreign researchers due to its strategic military position. In particular, the Crimean War played a role in keeping the Crimean naval bases from Western spies, an attitude that continued, and even strengthened, during the Soviet era. Despite its popularity as a summer vacation destination for workers, intelligentsia and party members, a large part of the peninsula's territory was closed to foreign visitors—and even to regular Soviet citizens. The era of virtual isolation came to an end in the early 1990s, as cooperation between local and foreign scientists opened up a new era in the study of Crimea's natural and cultural resources.

My contact with Crimea began soon after the area around Sevastopol was finally opened to foreigners in the years following the collapse of the Soviet Union. At that time, the Institute of Classical Archaeology (ICA) of the University of Texas, my doctoral *alma mater*, was invited to collaborate with local archaeologists in the survey and excavation of the ancient Greek city of Tauric Chersonesos, located in the middle of an area highly sensitive for its proximity to naval bases in Sevastopol. Professor Joseph Carter, the ICA director, invited me to join efforts to study the paleoenvironments of the region, along with one of my best friends, and former fellow graduate student, Paul Lehman.

Despite our focus on the Heraklean Peninsula—the southwestern corner of the Crimean Peninsula—my interest in Crimea's natural and cultural riches quickly extended to the entire territory of Crimea. In the development of my research I became acquainted with numerous local colleagues—scientists in geography, life sciences, ecology, history and archaeology—some of whom encouraged me to pursue further studies in the other parts of Crimea.

After publishing a number of articles in scientific journals and chapters in edited volumes (Cordova and Lehman 2003, 2005, 2006; Cordova, Rybak and Lehman 2003; Cordova 2000, 2007; and Cordova et al. 2011), I came to realize that it was time to bring together all the bibliographic, archival, pictorial, cartographic and field information collected during the previous 14 years and put them into a comprehensive volume. This book is the result of more than a decade of work in Crimea and a compilation of information gathered through exhaustive research in publications and reports in Russian and Ukrainian, some of which are difficult to obtain in the West.

The publication of this book has the following objective: to provide a broader audience with an idea of the natural and cultural wonders of the Crimean Peninsula by reviewing the main stages in the formation of its landscapes. In so doing, I review a variety of topics, from natural ones (biogeography, relict, endemic and rare species) to cultural ones (the evolution of the cultural landscape, the history of Crimean wine, and the peninsula's tourist potential), as well as aspects of preserving the natural and cultural heritage, encapsulated in the concept of conservation. Each of these inquiries corresponds to the themes of the book, discussed initially in Chapter 1 and arranged after as follows: Crimea's natural history (Chapters 2, 3, 4 and 10), environmental history (Chapters 5, 6, 7, and 9) and conservation (Chapters 10 and 11).

A high proportion of the data presented in this book has been compiled through field, bibliographic and archival research. Most of the bibliographic and archival research included extensive research in libraries and archives. A great deal of information was obtained from the library and archives at the Museum of Tauric Chersonesos in Sevastopol, the V.I. Vernadskiy National Library of Ukraine, the library at the Institute of Geography of the Ukrainian National Academy of Sciences, and the library at the Nikitsky Botanical Gardens in Yalta. Extensive research was also carried out at the New York Botanical Gardens and the libraries of the University of Wisconsin at Madison and the University of Minnesota, and the Botanical Institute at the University of Bern, Switzerland.

NOTES ON SOURCES AND USAGE

References in Russian and Ukrainian are transliterated using the Cyrillic transliteration keys in the next section. Translations are included in brackets for book titles and article titles only. If the reference is in Ukrainian, it will be indicated in parentheses at the end of the reference. The others will be assumed to be in Russian. The transliterated spelling of Russian and Ukrainian authors' names varies for articles published in western European languages. One example in the references of this book is Yanushevich, Janushevich and Yanushevitch, which are spellings for the same author. The same occurs for Wulf, whose direct transliteration from Russian is rendered as Vul'f, but in his publications in English or German is spelled Wulff. Therefore, when these cases occur in citation, the variations are included in brackets following the name of the author the first time it is cited in the text and in the list of references (e.g., Janushevich [= Yanushevich] 1979; Vul'f [= Wulff] 1937). Names of personalities known in the West are spelled according to standard English usage (e.g., Gorbachev, Khruschev).

Place names will generally follow the Cyrillic transliteration key in the next section, except for those of the most important cities, which are better known in the literature in the West (e.g., Kiev, Sevastopol, Simferopol, Yalta, Kerch, etc.), which are spelled according to Merriam-Webster's *Geographical Dictionary* (Merriam-Webster 1997). Another peculiar problem in Crimea and Ukraine is the transliteration standard used in English, whether it is from Russian or Ukrainian. Thus, Kiev and Kyiv are transliterations of the name of Ukraine's capital in Russian and Ukrainian, respectively. The same pattern is true for names of Tatar and Turkish origin, whose transliteration varies according to whether it comes from Turkish or Russian. For other place names, the same transliteration for authors' names in citations is used. Additionally, the translation of the names of some geographical features has been Anglicized given the difficulty of transferring the widely used adjectival form in the Russian language. Thus, *Baydarskaya Dolina* and *Sakskoye Ozero*, for example, are translated as the Baydar Valley and Lake Saki, respectively. But this does not apply to cases where the adjectival form is used in the English literature, as is the case of *Gerakleyskiy Poluostrov* (Heraklean Peninsula).

Use of the definite article as in "the Crimea" and "the Ukraine," very common in the literature of past decades, is not used here. Many colleagues, both Ukrainians and Americans, have expressed a concern that the use of the article implies that these two political territories were part of a larger political entity: that is, the Russian Empire and the Soviet Union. Ukraine has been an independent country since 1991, and Crimea is an autonomous republic within Ukraine. Furthermore, Ukrainian and Russian are two languages that have no use of the definite article. Therefore, it is more appropriate now to refer to them as Ukraine and Crimea.

Because the style for citing classical works is different, all references to works by ancient Greek and Roman geographers and historians are listed separately. The citation style follows the *Chicago Manual of Style*. Accordingly, the citation in the text is given by the name, book number, part and paragraph. Thus, Strabo (7, 4, 22) refers to Strabo's *Geography*, Book 7, Part 4, paragraph 22. In this case, Book 7 is contained in Volume III of the Loeb Classical Library. Therefore, the list of classical references will contain the information on this volume. References to websites are also listed separately, marked by the last year it was modified and the date it was accessed.

The number of plants and animals mentioned throughout the book is substantial. Normally, they are referred to by their scientific name, which is sometimes followed by the English common name or vice versa, as in the example *Taenatherium asperum* (medusahead rye). But for species that are constantly repeated, the common name is omitted throughout the rest of the chapter. Only in cases when the discussion is about particular aspects of taxonomy, is the full scientific name used, as in the example (*Taenatherium asperum* (Simonk.) Nevski). Only when reference is made to a group or genera is the common name used, as in the example oaks (the genus *Quercus*), pines (the genus *Pinus*).

Throughout the book, ages are often presented in years BP (before the present). In other cases, zeros are eliminated by using ka (thousand years) or Ma (million of years). Thus, 10 ka is 10,000 years, and 1 Ma is 1 million years. More specific dates, particularly in historical periods, are presented as BC or AD, for the case of years before the Common Era or after the Common Era, respectively. Non-calibrated radiocarbon dates are simply presented on illustrations as number of years plus or minus the uncertainty followed by RCYBP (radiocarbon years before the present), as in the example 2340 + / − 35 RCYBP. Calibrated dates are presented with the designation "cal" followed by either years before the present (BP) or calendar years (BC/AD):

e.g., cal 2530 BP or cal 580 BC. In some cases, calibrated dates may be presented as a time range (e.g., cal 2600–2300 BP).

The use of periods of climate events (climatostratigraphy) is based on the global Marine Isotopic Stages (MIS). In other cases, European equivalents or marine isotope stages (MIS) are used, e.g., the last interglacial also appears as Mikulino Interglacial (Russian), Eemian Interglacial (northern Europe), or MIS 5e. Refer to Figure 5.1 for the chronology of climatic events and their correlation with local archaeological periods. Although the terms glacial and interglacial are adjectives, they are used here as nouns, following the common usage in the scientific literature. For the Holocene, climatic events are referred to by regional stratigraphy, using Vyeklich et al. (1993) and Gerasimenko (2007). Refer to Figures 6.1 and 7.1 for climatic and archaeological periods used in this book.

RUSSIAN AND UKRAINIAN CYRILLIC SCRIPT TRANSLITERATION

Russian	Transliteration	Ukrainian	Transliteration
А а	A a	А а	A a
Б б	B b	Б б	B b
В в	V v	В в	V v
Г г	G g	Г г	H h
Д д	G g	Ґ, ґ	G g
Е е	Ye ye	Д д	D d
Ж ж	Zh zh	Е е	E e
З з	Z z	Є, є	Ye, ye
И и	I i	Ж ж	Zh zh
Й й	Y y (consonant)	З з	Z z
К к	K k	И и	Y y (short vowel)
Л л	L l	І і	I i
М м	M m	Ї ї	Yi yi
Н н	N n	Й й	Y y (consonant)
О о	O o	Ий	yy = y
П п	P p	К к	K k
Р р	R r	Л л	L l
С с	S s	М м	M m
Т т	T t	Н н	N n
У у	U u	О о	O o
Ф ф	F f	П п	P p
Х х	Kh kh	Р р	R r
Ц ц	Ts ts	С с	S s
Ч ч	Ch ch	Т т	T t
Ш ш	Sh sh	У у	U u

Russian	Transliteration	Ukrainian	Transliteration
Щ щ	Shch schh	Ф ф	F f
Ъ	“	Х х	Kh kh
Ы	Y y (short vowel)	Ц ц	Ts ts
ый	Yy = y (when at the end of the word)	Ч ч	Ch ch
Ь	“	Ш ш	Sh sh
		Ш ш	Shch schh
Э э	E e	ь	“
Ю ю	Yu yu	Э э	E e
Я я	Ya ya	Ю ю	Yu yu
		Я я	Ya ya

CHAPTER 1

CRIMEA AND THE BLACK SEA: THEMES IN ENVIRONMENTAL HISTORY

Crimea's geography

The Crimean Peninsula occupies a territory of approximately 27,000 km^2; an area about the size of the state of Massachusetts, or somewhere between that of Macedonia and Rwanda. Crimea is bounded to the west and south by the Black Sea, to the east and northeast by the Azov Sea, and to the north by the Sivash—a lagoonal system connected to the Azov Sea. The extreme latitudes are N 46° 10' 26" at the far north of the Perekop Isthmus, and N 43° 23' 11" at Cape Foros (Figure 1.1).

The territory of the Crimean Peninsula is composed of the plains, the piedmont and the mountains (see Figure 1.1). The summit of the mountains consists of a series of karstic plateaus known as yailas. The highest is the Roman Kosh (1,545 m), located in the Babugan Yaila in the central part of the Crimean Mountains. The mountains present an asymmetric topography, gentler to the north and steep on the south (Figure 1.2). The major river basins are located on the northern slopes of the mountains. The largest rivers are Salgir, Al'ma, Kacha, Bel'bek and Chyornaya. The streams of the southern slopes consist mainly of mountain torrents (Figure 1.2). The natural lakes of Crimea comprise saline shallow bodies of water scattered through the north and west of the plains and along the coasts of the Kerch Peninsula.

Modifications to the drainage system are seen through the numerous dams and canals. The most important of these canals is the North Crimean Canal (NCC) (Figure 1.2). Finished in the 1960s as part of Nikita Khrushchev's Virgin Lands Campaign, the NCC provides water for crops in an area naturally too dry or unpredictably moist for large-scale agriculture. The canal also serves as a water supply to towns, as well as being used for

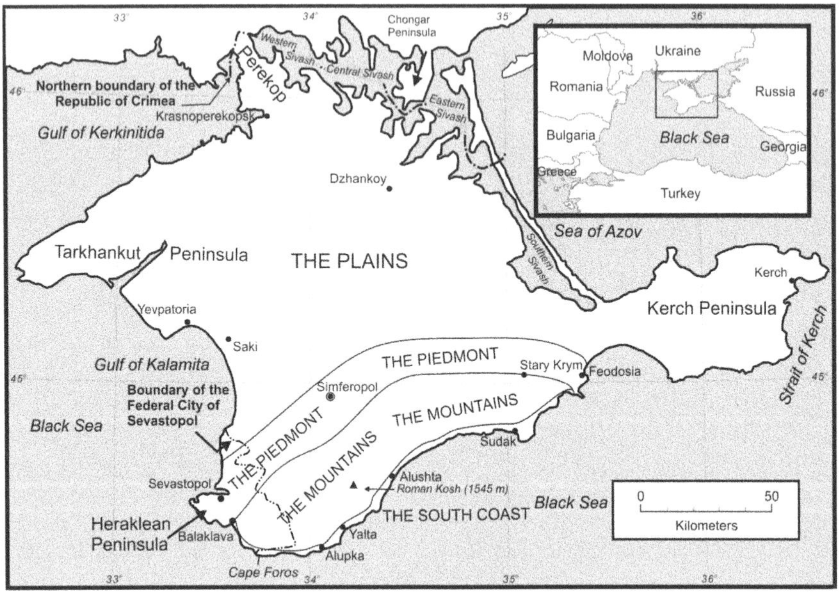

Figure 1.1: The Crimean Peninsula: Physiographic regions.

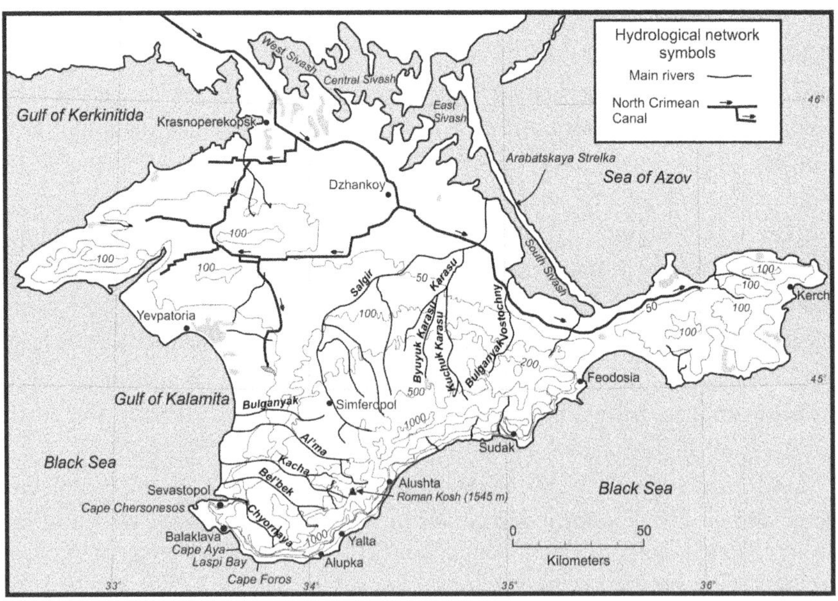

Figure 1.2: The Crimean Peninsula: Topography and hydrological network, including the main canals.

Figure 1.3: The North Crimean Canal (branch of the Dnieper Canal) near the town of Dzhankoy. Besides irrigation, it is used for recreation and fishing.

recreational activities (Figure 1.3). Although the NCC was an engineering marvel that had a practical economic purpose, it brought an end to the grasslands of the steppe. Since its completion, virtually all the steppe has been turned in to cropland.

Politically, since its annexation to the Russian Federation in March 2014, the Crimean Peninsula comprises the District of Crimea, which in turn is divided into the Republic of Crimea and the Federal City of Sevastopol (Figure 1.1). Although a subject of Russia, the Republic of Crimea has its own constitution and flag. The Federal City of Sevastopol is one of three cities with that status in the Russian Federation (the other cities being Moscow and Saint Petersburg).

The Republic of Crimea (the Autonomous Republic of Crimea under former Ukrainian rule) had a total population of 1,973,000 (as of 2007). Its capital city, Simferopol (340,000 inhabitants) is located on the piedmont on the Salgir River. Other major cities are Kerch (157,000), Yevpatoria (103,244), Feodosia (97,721) and Yalta (80,552). The Federal City of Sevastopol (the Municipality of Sevastopol under Ukrainian rule) had a population of 379,200 inhabitants (as of 2007). Its major cities are Sevastopol (342,000), Balaklava (30,000), and Inkerman (10,472). The total population of the Crimean Peninsula is 2,352,200, with a population density of 89 people per square kilometer. The main ethnic groups in the entire peninsula are Russian, Ukrainian and Crimean Tatar, followed by smaller groups such as Armenians, Jews, Byelorussians, Bulgarians, Greeks and Italians.

Russian is the most widespread language, although Ukrainian is also spoken by a minority. However, Crimean Tartar, a language of the Turkic branch of the Ural-Altaic family, is spoken among the growing Tartar population of Crimea. The most widespread religion is the Orthodox Christian faith, which is practised by a sizeable number of the Slavic inhabitants of the peninsula. Sunni Islam is practised mainly by the Tartar population. Additionally, Judaism and various Christian sects are practised by much smaller groups.

Biogeography and environmental history

This book presents and discusses aspects of Crimea's natural world in a dynamic and historical context. In this endeavor, the topic of the book—the making of Crimea—draws on disciplines such as geology, climatology, geomorphology, Quaternary science, biogeography and environmental history. But given the specific themes of the book (discussed in this chapter) the emphasis is on biogeography and environmental history.

Biogeography, or the study of spatial distribution and dynamics of living organisms, includes analysis of topics such as endemic and relict species, the evolution of communities, populations, biomes, and their strong links to geological history and climate change. The interaction of these biogeographical aspects of the peninsula with human societies is what constitutes the environmental history. In turn, biogeography and environmental history help us understand the modern landscape of Crimea; to comprehend complex environmental processes such as biodiversity and conservation issues; the potential for tourism; and offer guidelines for scientific studies on nature, history and archaeology.

As a biogeographically focused work, this book revisits many of the hypotheses and ideas put forward by Soviet biogeographers regarding the origin of Crimea's flora and ecosystems. The most frequently discussed of these hypotheses are the ones that try to explain the origin of the Mediterranean vegetation in southern Crimea. My approach to the problem is based on my own palynological and biogeographic research, taking into account the paleoclimatic, geological and ecological research carried out in recent years in the Black Sea region. The topic involves aspects of European glacial *refugia* of trees and boreal relict flora in the mountains. Recently, palynological studies by Ukrainian palynologist Natalia P. Gerasimenko have revealed more information on full-glacial vegetation in Crimea (Gerasimenko 2007). These studies have also been complemented with research on faunal remains associated with Neanderthal and modern human occupations during the Middle and Upper Paleolithic, namely, the last interglacial–glacial cycle (Chabai, Marks and Monigal 1999, 2004).

Environmental history, unlike natural history, encompasses the presence of humans as modifiers and often builders of the environment. It focuses on human–environmental relations. Thus, for Crimea, an environmental history should start with the time when bands of hominins inhabited the peninsula during the Pleistocene. The Neanderthals (*Homo sapiens neanderthalensis*), often referred to as cave men, inhabited Crimea from at least since the last interglacial period (128,000–115,000 years ago) until their demise and replacement by their successors, the anatomically modern humans (*Homo sapiens sapiens*), sometime between 40,000 and 30,000 years ago. Afterwards, human history becomes more and more relevant to environmental history, since modifications of the environment increase, leading to the introduction of agriculture and further cultural developments.

In theory, there were hominids in Eurasia before the Neanderthals, but evidence of their presence in Crimea exists only in very few places. If such occupations were once widespread they are not visible or have been wiped out by erosion and other natural processes. The environmental history here, therefore, encompasses that part of prehistory that in Crimea is better known, which extends from the last interglacial period, through the last glaciation, and the postglacial period known as the Holocene.

Archaeology and history also have a decisive role in the reconstruction of environmental history. The archaeological record merges with the historical record some time before the beginning of the Common Era. Documents referring to the Crimean Peninsula appear together with Greek and Roman historians and geographers, who apparently wrote very little about it. Being on the edge of empires and classical influence, the northern Black Sea shores are known less than the Mediterranean core region but, as time passes, more descriptions and references appear, particularly during and after Byzantine rule of the peninsula (roughly from the third to the fifteenth centuries AD). Therefore, our understanding of human–environment relations across the various historical periods is still in the making and is constantly being updated, with more research projects currently underway.

The multiple aspects of biogeography and environmental history of any territory, regardless of its territorial size, are difficult to compress into a single volume. Furthermore, it is difficult to separate them as topics that are exclusively biogeographical or exclusively environmental history. This being so, the topics treated in this book follow seven main themes, described below.

Isolation and connection

In his novel, *The Island of Crimea*, Vassily Aksyonov, a dissident writer of the former Soviet Union, portrays Crimea as an island, whose geographical conditions allowed the White Russian Army to resist the Bolshevik conquest and to form a Russian Republic. In the novel, in the late 1970s the Republic of Crimea has evolved into a country in the sphere of the capitalist West. The idea of an island as portrayed by the novel is not difficult to imagine: just look at the map of Crimea. It is possible to see that the connection to the mainland is just a narrow isthmus, the Pekerop, which separates the Gulf of Kerkinitida from the Sivash lagoonal system (Figure 1.1).

The Perekop (isthmus, in Russian) has a minimum width of 2 km, an approximate length of 7 km, and a maximum elevation of 20 m above sea level. Today, the Perekop is not the only access to Crimea from the mainland: a road and a railway line connect the mainland with Crimea on the central part of the Sivash via the Chongar Peninsula through bridges across a strait no wider than 100 m and no deeper than 1 m. Another bridge connects the mainland with Arabatskaya Strelka on the eastern part of the Sivash. However, the traditional and historical land access to Crimea has been the Perekop.

The idea of insularity depicted in Aksyonov's novel has a strong parallel in writings about Crimea's natural history, biogeography, geology and archaeology. During the warm stages, or interglacials, the sea rose to levels similar to the modern one or even higher, suggesting possible flooding of the lowlands of the northern part of the peninsula, namely the Perekop and Sivash areas. Total separation from the mainland, created by a sea on what is now the Sivash region and the Perekop, has been proposed by archaeologists (e.g., Lazukov et al. 1981; Richter 2005; Chabai 2007) for the last interglacial (130,000–115,000 years ago), but its existence is doubtful as no stratigraphic evidence in this region has yet confirmed the existence of a strait connecting the Black Sea with the Sea of Azov (see discussion in Chapter 5).

Even if the alleged isolation did not occur, which is the opinion of the author, Crimea would still have been a functional island in the biogeographic sense. A harsh, hostile environment created by high salinization and unpalatable vegetation in northern Crimea would also act as a barrier similar to a sea. The picture is clear even in the historic landscape of northern Crimea today, so that the biogeographer Lev S. Berg (1950) considered the extensive salty dry steppes around the Sivash to act as a barrier to the movement of fauna and flora, thus adding to the biogeographic isolation of Crimea.

Yet the idea of isolation is not complete if it is not contrasted with the idea of connectedness, i.e., the suggestion that Crimea may have been connected with regions around the Black Sea other than through its present connection with the Ukrainian mainland. Like isolation, connectedness has also stirred lengthy discussions among naturalists and academics, and particularly geologists and botanists, who have always thought in terms of bridges between Crimea, the Caucasus, the Balkans and Asia Minor (modern territory of Turkey). This idea evolved into what is called the Pontida Hypothesis, which alludes to the existence of a former land bridge between the Crimean Mountains, the Caucasus and Asia Minor. As explained in Chapter 10, this idea has been dismissed by geologists and botanists themselves, and replaced with more modern hypotheses concerning changes in sea level in tandem with transitions between glacials and interglacials. Linked to the issue of isolation, connectedness and land bridges, are the concepts of biodiversity, endemisms and relicts, which have also been a hot topic in Crimean biogeography and conservation.

Biodiversity, or the number of species per area, is another biogeographic issue that accentuates the insularity–connectedness relation. Areas of high biodiversity in the world have been designated biodiversity hot spots, of which the Caucasus is one (Myers et al. 2000). Despite biogeographic affiliations with the Caucasus, the Crimean Mountains are not part of the Caucasian Biodiversity Hotspot. Species diversity in Crimea, although high by world standards, is low compared with that of the Caucasus (Biodiversity Support Program 1999). In view of this, and the previously mentioned issue on insularity, I offer in Chapter 10 an alternative interpretation of the biogeographic issues of biodiversity, endemics and relicts by characterizing the Crimean Mountains as a "functional island."

The influence of the Black Sea

The Black Sea has had a tremendous influence on practically all aspects of Crimea's natural and cultural history. Firstly, the Black Sea influences the climate of the peninsula by providing a source of moisture and by preventing temperatures from attaining high and low extremes. Secondly, the Black Sea has been the main factor in the process of connection/isolation with the mainland. Finally, the Black Sea has been a carrier of natural and cultural agents in and out of the peninsula. For this reason, it is difficult to talk about aspects of biogeography and environmental history without acknowledging the surrounding sea as an important factor in shaping the Crimean landscape.

The origins of the names given to the Black Sea in antiquity are shrouded in mystery. While early classical works refer to it as *Pontus Axenos* (unwelcoming sea), presumably an allusion to the difficulties of navigation and to the storms, in late antiquity the name changed to *Pontus Euxinus* (the hospitable sea) (King 2004). The name "Black Sea" is as dubious as its predecessors. The name comes from the translation of the Turkish *Kara Deniz*—literally, Black Sea. Interestingly, the Turks refer to the Mediterranean Sea as *Ak Deniz* (White Sea). The meaning of these two names suggests that their country is between two remarkably different seas. Modern Turks refer to this difference in terms of weather: stormy and dark skies characterize the Black Sea, while nicer and more stable weather characterizes the Mediterranean. However, Charles King (2004) proposes that the early Greeks adopted the Iranian name *Axenosi*, from the Iranian word *Axeinos* (dark), which seems to suggest that the illusion to the blackness or darkness of the modern name originated in antiquity.

The Greek geographer Strabo (64 BC–24 AD) wrote extensively on the Black Sea. Strabo was originally from Amaseia (modern Amasya), near the northern coast of modern-day Turkey south of Sinop, which probably left him with a knowledge of many of the traits of the Black Sea from contact with sailors and fishermen. Although a geographer, Strabo is not particularly characterized as a traveler; most of his books are based on secondary sources, particularly from the Library of Alexandria (Hazel 2002). Therefore, the likelihood that he had ever traveled in Crimea and the northern shores of the Black Sea seem slim.

One of the most interesting features of the Black Sea and Crimea that Strabo describes are the sea currents and their use by ancient Greek sailors (Figure 1.4). The most relevant fact related by Strabo on sea currents and sailing is his reference to the Megarian route. Named after the city of Megara, in the Greek mainland, the Megarians were an offshoot of the Dorian Greeks,

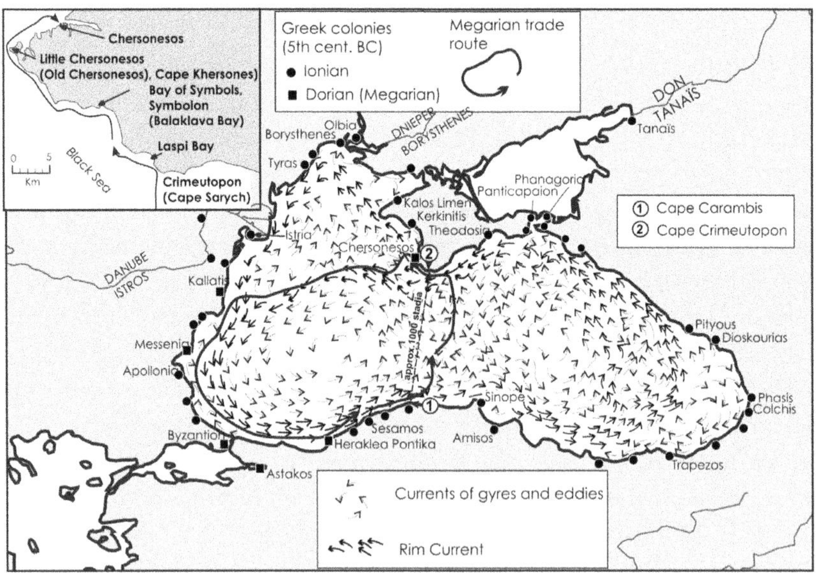

Figure 1.4: The Black Sea currents, Strabo's Megarian route, and Greek colonies of the First Millennium BC.

who were successful in establishing cities and trade routes in the western half of the Black Sea basin (Hind 1998).

Strabo (2, 5, 22; 7, 4, 3) was the first to observe that a 1,000 stadia long line between Cape Crimeutopon on the southern Crimean coast, and Cape Carambis on the northern coast of modern Turkey, effectively partitioned the current into two seas (Figure 1.4). It is perhaps obvious that he also observed the two main gyres, then generating two currents, one in the west with counter-clockwise rotation and one in the east with clockwise rotation. The gear effect of the two gyres results in the counter-clockwise Rim Current, which was also important for cabotage navigation (Zolotaryev 1979). Thus, the Megarian route of colonization simply follows the Rim Current for the most part (Hind 1998). One example of this is the chain colonization of Dorian Greek cities, beginning with the mother town of Heraklea Pontika (present-day Ereğli) (Figure 1.4).

The importance of the Rim Current goes beyond its significance in ancient navigation. The current seems to have had a role in the possible migration of plant species from Asia Minor and Crimea (Cordova 2007). A collection of floating garbage items along the shores between Laspi Bay and Cape Foros in the spring 2001 produced an interesting count of materials collected from different parts of the Black Sea, suggesting that many products originated in Turkey (Cordova 2007). The late Crimean geomorphologist, Alexander A. Klyukin, commented that, after storms, drift currents along the coast often deposit tree stumps. He observed the rocks trapped in the root systems of these stumps. On several occasions he found rocks that do not exist in Crimea. This suggested transportation via sea currents (A. A. Klyukin, personal communication).

The Black Sea current system results from the intensification of winds associated with cyclones together with the effect of the Crimean Peninsula, which acts as a promontory dividing the currents into two main gyres, east and west (Saenko, Knysh and Korotaev 2000). In turn, the convergence of the northward currents of the two gyres at the south coast of Crimea has an impact on the temperatures of the water along the south coast, particularly in winter, when the currents intensify as cyclonic activity increases (Saenko, Knysh and Korotaev 2000). Additionally, the warmth brought by the cyclones from the Mediterranean keeps surface water temperatures above freezing level (Kostopoulou and Jones 2007). The fact that the water is deeper here than for any other offshore area of Crimea and that there are no major rivers emptying into the sea, are two additional factors that make the coast of Crimea from the mouth of Sevastopol Bay to Theodosia an ice-free coast in winter.

The only natural harbors along this ice-free coast are Sevastopol Bay, Balaklava Bay, Sudak Bay and Caffa Bay (present-day Feodosia). Ports had existed in these place throughout antiquity and the Middle Ages, but the largest of them and best fit for a large-scale navy was Sevastopol Bay. During his travels through the southern Russian province in the 1830s and 1840s, Captain Edmund Spencer (1855) noticed the many strengths and weaknesses of the Russian ports along the shores of the Black Sea and the Azov Sea. Among the downsides of the major ports (e.g., Rostov-na-Donu and Odessa) was ice forming during the winter months. The only exception was Sevastopol, which being on the southwestern coast of Crimea was ice free. As a strategist, Captain Spencer wanted to get the point across that Sevastopol, the contested port in the Crimean War (1853–4), was vital for the all-season movements of the Russian navy.

The influence of sea currents on Crimea's natural and cultural history is overshadowed by another dynamic aspect of the Black Sea: changes in sea level. The topic, however, remains a controversial one, since researchers still disagree on the magnitude and timing of these changes. The most controversial topic regarding sea-level change in this region is the postglacial refilling of the Black Sea basin, often dubbed the Black Sea Flood. Two oceanographers from Columbia University, William Ryan and Walter Pittman (1998), proposed a catastrophic scenario in which the Mediterranean Sea rose above the lowest point of the Dardanelles Sea of the Marmara–Bosporus passage, consequently spilling over into the Black Sea basin—then occupied by a freshwater lake some 120 m below its present level. According to William Ryan and Walter Pitman, this was a catastrophic event that created an enormous waterfall and a rapid refilling of the Black Sea to its modern levels. Because of its catastrophic nature, the authors dubbed this event "Noah's Flood." The authors went further to propose that the event may have been recorded by cultural groups and passed down generation after generation until it became known to several religions and beliefs in the area as a catastrophic flood event. Radiocarbon dating of sediments put this event at sometime around 7,500 years ago.

William Ryan and Walter Pitman's theory was challenged by a number of marine geologists. Most critical of their theory were Ali Aksu et al. (2002), who claimed that the process happened by the Black Sea flooding into the Mediterranean, and by Andrei Chepalyga (2007), who gives importance to the waters of the Caspian Sea, which spilled over into the Black Sea basin through the Manych Strait. Despite the different scenarios proposed, the filling-up of the Black Sea basin is undeniable. As global temperatures rose, sea levels worldwide also rose as a consequence of the melting ice sheets. This rise of the global oceans was soon felt in the smaller seas, which in turn reconnected with the global ocean.

The controversy surrounding the mechanism of Mediterranean–Black Sea reconnection centers on the timing and processes of the reconnection of the Black Sea and the Mediterranean after the Ice Age and it prompted a meeting sponsored by Columbia University in 2003, the outcome of which was a published volume by Yanko-Hombach et al. (2007). In 2005, a scientific team sponsored by UNESCO's International Geological Correlation Program 512 (IGCP-512) began to organize meetings to discuss the results of scientific research in the disciplines of geology, oceanography, archaeology and other subjects, in order to study in depth the connections between the Mediterranean, the Black Sea, and even the Caspian Sea basins. Regardless of modes and timing of the postglacial refilling of the Black Sea basin, it would be interesting to evaluate how the postglacial sea-level rise affected the overall natural and cultural aspects of Crimean Peninsula.

Through glacials and interglacials

Unlike other mountainous areas of Europe and the Caucasus, glaciers did not develop in the Crimean Mountains during the Pleistocene glacial periods. However, faunal and floral remains, soils and sediments, and other paleoenvironmental records provide evidence of profound landscape changes. In the coldest stages, the sustained dry-cold conditions of the plains lead to the accumulation of loess (dust) in a manner similar to the plains in mainland Ukraine. By contrast, the mountains and the piedmont sustained a steppic environment with some boreal trees and pines and with fauna typical of the Eurasian steppe and some sub-Arctic species. But evidence exists of broadleaf temperate trees surviving in the valleys and canyons of the mountains and probably on the southern shores of the peninsula, hence creating glacial *refugia* (see data and discussions in Chapters 5 and 10).

The biogeographic concept of glacial *refugia* (plural: *refugium*) implies that as the ice advanced on the northern and central parts of Europe trees migrated to warmer parts of the continent, where they survived the coldest stages of the glaciation. Glacial tree *refugia* are often found in those areas of Europe where the harsh glacial climatic conditions allowed temperate and boreal tree species to survive throughout cold stages, often in the Mediterranean regions or on the southern side of mountain systems (Comes and Kadereit 1998; Leroy and Arpe 2007). Conversely, interglacial periods represented the re-colonization of deglaciated areas by tree species. In the case of Crimea, however, the interglacial cycles represented the re-colonization of thermophilic species, such as those of Mediterranean origin.

The south coast and the Mediterraneanization of Crimea

Dubbed the "Russian Riviera" during imperial times, the south coast of Crimea attracted both nobility and artists, who saw in Crimea a subtropical paradise that contrasted with the rest of the empire, where harsh winters and torrid summers were the norm. For the same reason, during the Soviet period the southern coast became one of the most important tourist destinations for workers, intelligentsia and the communist party leadership.

The mild Sub-Mediterranean climate on the south coast, together with the beauty of those coasts (normally seen as equal to those of Italy and the French Riviera) have drawn the attention of geographers and government officials. Soviet leaders not only saw Crimea as the place to spend holidays, as it had been for their noble predecessors in imperial times, but also as the place to establish sanatoria (health resorts) for workers and the intelligentsia. Today it remains a major tourist destination, mostly frequented by Ukrainians, Russians and citizens of other ex-Soviet republics. The dry and warm climate, the warm water temperatures and the beauty surrounding the sea are indeed the main attraction (Figure 1.5). The warmth of Crimea's climate also attracted naturalists and agronomists, and the renowned Nikitsky Botanical Gardens (founded in 1812), became the leader in research geared towards introducing thermophilous vegetation, for ornamental as well as agricultural purposes.

Vineyards are an important part of the landscape, not to mention the large number of plants native to the Mediterranean region, as well as introduced cultivated and ornamental plants, ranging from olives and fig trees to Italian cypresses. A number of the native flora of the southern coast of Crimea constitutes elements typical of the Mediterranean biome. The majority of such elements are not found farther north, having southern Crimea as the

Figure 1.5: View of the Crimean south coast from Maloryechyenskoye on a crowded beach. In the background is Ayu-Dag, or Bear Mountain, the southern slope of the Crimean Mountains, and the Babugan Yaila with typical summer cloud mantle.

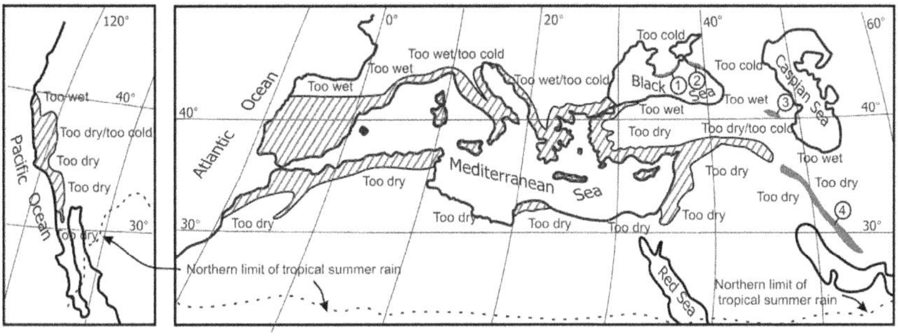

Figure 1.6: Schematic comparison of latitudinal location of Mediterranean climate regions of the Northern Hemisphere and exclaves of Mediterranean vegetation: (1) southern Crimea, (2) Tuapse-Novorossiysk region, (3) southern Armenia, and (4) western Iran.

northernmost location of their geographic distribution. This is the case for *Pistacia mutica*, *Cistus tauricus*, *Ruscus ponticus*, *Arbutus unedo* and *Pyracantha coccinea*, among others (Vodop'yanova 1986).

The warm climate and the presence of Mediterranean floristic elements and soil types along the south coast of Crimea has been a topic that has intrigued both naturalists and geographers. This region is not only geographically detached from the Mediterranean biome: it is also located a latitude too high for Mediterranean vegetation (Figure 1.6). Peter Simon Pallas (1812) indicated that most of the Mediterranean plant species of the south coast of Crimea were brought and naturalized by humans. Although, today, a large number of Mediterranean plants are naturalized species, paleobotanical evidence shows that a large number of them are native (Cordova and Lehman 2005; Cordova 2007).

Most of the theories proposed to explain the origin of Crimea's native Mediterranean vegetation point to remains of a disjunctive relict territory dating back to the Miocene and Pliocene, which is based in the presence of relicts and endemic plants (Didukh 1992). However, this theory raises another issue: How did Mediterranean flora in Crimea survive the glacial periods of the Pleistocene? The question as to whether Mediterranean vegetation survived the glacial stages in local *refugia* or retreated to southern latitudes is still one of the major problems in understanding the continuum of ecosystems of the southern Crimean Coast.

In view of the issue posed above, the questions raised in this book are: Is southern Crimea truly an exclave of the core Mediterranean region? How different is southern Crimea, naturally and culturally, from the Mediterranean region? These questions can be answered from two different standpoints: the history of native and naturalized Mediterranean vegetation, and the processes of adaptation involved in the introduction of Mediterranean agricultural elements in the late Holocene.

The main characteristics of a Mediterranean climate are mild winter temperatures and pronounced summer drought, which the Köppen classification of climates defines as Csa or Csb. However, some of the typical Mediterranean plant species do not always stay within the range delimited by the Köppen criteria (Rikli 1943). However, there are numerous criteria to define the limits of the Mediterranean biome, ranging from those purely climatic (e.g., Emberger 1955; Daget and David 1982; Le Houérou 2004) to biogeographic (e.g., Rikli 1943; Zohary 1971, Walter 1974) and even cultural (e.g., King 1997).

Due to its strong attachment to the Mediterranean type of climate, particularly in terms of temperature extremes, the limit of olive cultivation has often been used as a limit for the Mediterranean type of climate (King 1997). However, olive (*Olea europaea europaea*) is now a domesticated and human-propagated species found in many continents and a diversity of climatic conditions in subtropical latitudes (20–30° N and S). But in Crimea winters are too cold for the development of olive cultivation. Although a few olive trees were reported by travelers, olive pollen has never been found in any deposits in Crimea (Cordova and Lehman 2003, 2005). In the twentieth century, experiments led to the successful production of olives in the Yalta area, the warmest part of the peninsula (Vul'f, Kalayda and Plotnistkiy 1916). But for reasons other than climatic limitations, olive never became an important crop.

Douguedroit and Zimina (1987) compared several climatic traits of Crimea's south coast with selected localities in the European Mediterranean region. These traits include a summer drought and mean winter minimum temperatures slightly above freezing. The study shows that compared with the Mediterranean localities, the southern Crimean coast has a shorter summer drought and lower minimum winter temperatures at sea level. Their study found that the closest climate was to be found in southern France, slightly north of the Mediterranean climate region (Douguedroit and Zimina 1987). In fact, by latitude southern Crimea is only one degree above the northern limit of the Mediterranean in northern Italy and southern France (Figure 1.6). By comparison, the northern limit of the Mediterranean climate zone of North America's south coast is much lower in latitude. If Crimea's south coast latitude is traced horizontally to North America, the line will be at the latitude of northern Oregon, southern Minnesota and central Wisconsin.

In addition to southern Crimea, other territories in southwest Asia have been identified as exclaves of Mediterranean vegetation (Figure 1.6). These exclaves include a narrow strip along the coast of the western Caucasus, just south of Novorossiysk (Didukh, Litvinskaya and Novosad 1990), and two regions in southwest Asia: in southern Armenia (Walter 1974) and the western Zagros Mountains of Iran (Zohary 1971) (Figure 1.6). Although not part of the Mediterranean climate region, these exclaves are disjunctive territories of a once larger Mediterranean biome at the end of the Tertiary (Raven 1971; Zohary 1971; Didukh, Litvinskaya and Novosad 1990).

The south coast of Crimea not only presents traits of a Mediterranean type of climate, but it is also a relict territory of a larger Mediterranean bioregion, and an area with cultural influences from the Mediterranean, which is a process epitomized by the introduction of Mediterranean crops and rural management patterns. This process of introduction has been facilitated by Crimea's Sub-Mediterranean climatic and biogeographic traits, which constitute a process referred to as the Mediterraneanization of Crimea (Chapter 12).

Forests and steppes: the natural and cultural duality

In terms of natural and cultural aspects, the Crimean Peninsula is two worlds in one: the mountain region and the steppe. The mountain region, although only a third of the peninsula, holds the largest number of plant and animal species, and the largest number of endemic, rare and relict species (Biodiversity Support Program 1999). By contrast, the steppe is characterized by an immense flat and rolling plain dominated by herbaceous vegetation, although the monotony of the landscape is broken by diverse vegetation units of grassland and mixed-herb vegetation.

In terms of affinities of floristic elements, the mountains and steppe of Crimea present a dichotomy. While the mountains have floristic elements akin to those of the Caucasus, Balkans and Asia Minor, the steppes have elements typical of the Ukrainian and Russian steppes (Didukh 1992). No sharp boundary exists between the two. The forest–steppe, located mainly along the piedmont, constitutes the ecotone between these two units.

The mountain–steppe dichotomy goes beyond floristic elements to become a cultural one. Over millennia the peoples of both regions have been remarkably different. Since the Neolithic, the mountains have experienced numerous changes as agriculture and pastoralism have provided the way of life of sedentary and semi-sedentary people. The steppes remained as the land of nomadic peoples until recent centuries.

Diverse ethnic groups in a diverse landscape

The human populations that arrived and stayed in Crimea during the past three millennia have forged the modern ethnic, social and economic mosaic and landscapes of the peninsula. These groups range from the highly mobile nomads, some of whom were warring tribes, often depending on horses for their mobility, to highly organized groups that formed the Greek city-states. The groups included peoples speaking languages affiliated to different families: Taurians, Cimmerians and Scythians spoke languages in the Indo-Iranian Group; Greeks, Romans, Goths, Venetians, Genoese, Germans, Russians and Ukrainians spoke languages in the Indo-European group; Alans, Huns, Kazhars, Tatars and Ottomans spoke languages of the Ural–Altaic group.

The major Greek and Byzantine influence, as well as Jewish, Gothic, Turkic and Italian, remains in the southern mountainous areas. The influence can be seen in the ruins of fortresses and cities in the mountains, as well as in the customs and land-use practices in the countryside. In the steppe, fewer monumental legacies exist of the nomadic tribes that inhabited the country. One of them, however, is the *kurgany* (singular: *kurgan*), or mortuary mounds. But not being sedentary peoples there is not much left behind.

Developments following the annexation of Crimea by the Russian Empire in 1783 began the process of cultural change in the Crimean plains. First, the most important cultural development was that the Slavic element appeared in the picture as Russian and Ukrainians moved to the peninsula, both to mountains and steppe. Second, the agricultural settlement of the plains that culminated with the construction of the Dnieper canal and the breaking of the land for agriculture changed things completely. Unlike the mountains, where ancient influences persist, the original natural and cultural aspects of the steppe are now ancient history.

Legacies exist, not only as features left on ruins and other features in the landscape, but also in land-use practices inherited from the past. One of the best examples is the cultivation of grapes, which was introduced by the Greeks. Another land-use practice is the cultivation of nuts, particularly the walnut, which although attributed to the Greeks was more common in the Byzantine period (see Cordova and Lehman 2003). The cultivation of other Mediterranean products is also visible in many areas, particularly essential oils, fruits, pulses, etc. A blending of influences is also not uncommon: a Russian-style dacha (field house) might be surrounded by a typical Mediterranean orchard with fig trees, grape vines and a mixture of herbs such as mint, thyme and oregano.

Pastoralism was curbed during the twentieth century but it has recently been revived as Crimean Tatars in exile have begun to return. One example is the Baydar Valley, in the western sector of the Crimean Mountains, where a growing Tatar population is reverting land-use to forms seen in previous centuries. This includes the management of pastures, traditional farming and, more generally, the presence of cultural features such as mosques and other monuments.

Place names (or toponyms) reflect the cultural diversity and the historical influences on the landscape. The majority of place names today are of Tatar and Turkish origin, despite the fact that, throughout the past 200 years, names have changed back and forth between Russian and Tatar (Byelyanskiy, Lyezina and Supyeranskaya 1998). A number of non-Tatar names include Russian names or the russified version of the original name, whatever the language. Greek place names survive in the peninsula, particularly on the west and south coasts. Some of the Greek names, however, have been modified into Turkicized versions, as is the case of Ai Petri and Cape Aya, where the word *agia* (Greek for saint or holy) has been modified to Ai or Aya (Byelyanskiy, Lyezina and Supyeranskaya 1998). Other Greek names survive with the Russian spelling (e.g., Feodosia instead of Theodosia). Some Tatar names were replaced by Russian and Soviet names, some of which have recently been reverting to their old names. For example, a village in the mountains appears in some maps as Kuybishevo, but the road signs begin to show its original name: Albat. In recent years the trend has been to change the spelling of names to Ukrainian, which has been received with discontent among the largely Russian-speaking population of Crimea.

The historical change in place names is best reflected in the name of the peninsula itself. In antiquity the Crimean Peninsula was known as Tauria, or the land of the Tauri, the native tribe of the mountains at the time of the first contact with the Greek world. At some point the peninsula was also known as Gothia, given the establishment of the Goths there in the fifth century BC (Vasiliev 1936). After the takeover by Tatar-Mongol groups it was also known as Tartary. The name Crimea was not used until the Late Middle Ages, and although its origin is still debated by historians (Andryeyev 1997), it originated from Krym, the first Tatar capital, which was situated in the same location of the town known today as Stary Krym (or Old Krym) (Figure 1.1).

Tauria survives only in the scientific names of plants and animals (e.g., *Cistus tauricus* and *Podarcis tauricus*), archaeological sites (Tauric Chersonesos) and in the name of Crimea's largest university (Tauria National State University). The name Taurida (the Russian version of Tauria) prevailed for decades in the administrative region of Tavricheskaya Guberniya (Tauria Governorate), the name of the province that included Crimea and an area of the mainland, encompassing the part that was annexed to Russia in 1783. In later years, however, changes in the administrative division of the country brought the name *Krym* back to the peninsula, ending in *Krymskaya Oblast'* (Crimea District) during the Soviet period. Today, the official name of Crimea is the District of Crimea, with its subdivision, the Republic of Crimea.

Place names are of great value for the study of the environment and its perception by people. They not only reflect cultural aspects and perceptions of the environment, but also indicate location of resources, some of which may not exist anymore. This aspect was pointed out by the naturalist Peter Simon Pallas at the end of the eighteenth century. Although a naturalist, his work is of great relevance to the archaeology and history of Crimea. Needless to say, place names are a potential resource for an environmental history of the Crimean Peninsula.

Landscape transformation and conservation

The natural beauty of the landscape, together with its mild climate, has always been a resource for Crimea, particularly for the development of tourism. But human activity has continuously put pressure on the floral, faunal and geological resources that constitute that natural beauty. The fact that a number of rare and endemic floral and faunal species are threatened by human activities and climate change is an issue that worries environmentalists and scientists alike. Along with the threat to endangered species, the loss of biodiversity problem has worsened (Biodiversity Research Program 1999). But the need for protection, generally met through the creation of reserves, has faced many challenges. Although the problem is much discussed today, attempts at conservation and protection of species have a long history.

Knowledge of biogeographic traits such as endemic, relict and endangered species and their distribution has become crucial in establishing a base for the protection of species and ecosystems and it for this reason that the issue of nature conservation forms an indispensable part of this book. A knowledge of nature and environmental history are also critical for our understanding of this issue. In part, the idea of this book is to make available to the wider world the background and current status of research and action in the area of nature conservation.

CHAPTER 2

THE HISTORY OF ENVIRONMENTAL KNOWLEDGE

Early geographers and travelers

Crimea was known to the Graeco-Roman world as Tauria, named after one of its native groups, the Tauri or Taurians. References to its geography, natural properties and ethnic groups appeared in the works of Hecateus of Miletus (550–476 BC), Herodotus (484–425 BC), Theophrastus (371–287 BC), Strabo (64 BC–24 AD), Pliny the Elder (25–79 AD), Flavius Arrianus (86–160 AD), and Claudius Ptolemy (90–168 AD). None of these ancient writers devoted specific works to Tauria; their works are broad historical and/or geographical compendia about the known world of the time. Although their references to Crimea are very general, they provide broad descriptions of the landscape and its peoples. Of all the ancient writings, the most useful for surveying an environmental history of ancient Crimea are those by Herodotus (*Histories*) and Strabo (*Geographies*) and, to a lesser extent, Pliny (*Natural History*).

Written in the third century BC, Herodotus' *Histories* provides ample information on the ethnic groups inhabiting the regions surrounding the northern part of the Black Sea, particularly the Scythians (Herodotus 4, 2–103). Their territory, Scythia, includes mainly the southern part of the Ukrainian steppe and the plains region of Crimea. Herodotus' descriptions of Scythia include aspects related to the hard climate and rural activities, as well as certain conditions of the steppes. In relation to the climate of the Scythian steppes, Herodotus concludes that it is too cold and dry for farming but not bad for raising livestock (Herodotus 4, 28). Although their traditional lifestyle was nomadic pastoralism, in later times in Crimea the Scythians became sedentary (see Chapter 7).

Another interesting aspect of the steppe landscape mentioned by Herodotus is the "feathers flying in the steppe" (Herodotus 4, 7 and 4, 30).

It is possible that he is referring to the *Stipa* species of grasses, which are known as the feather grasses, or *kovyl'* in Russian. Several species of *Stipa* in fact are the dominant grasses of the True Steppe (see the vegetation description in Chapter 4).

In his *Geographies*, written in the first century BC, Strabo provides more information on natural features and the environment than any other account of Crimea in antiquity. Most of his descriptions of Tauria appear in Books 2 and 7, which contain ample and detailed references to coastal features and the territories of the Greek colonies at the time, particularly Chersonesos and Panticapaion, at the present location of Sevastopol and Kerch, respectively. He wrote about sea currents and sailing itineraries in the Black Sea with relatively good detail (Figure 1.4).

Strabo provides various references to features in the interior, such as lakes, rivers and mountains, although with a level of accuracy and detail that would create confusion and disappointment among modern scholars. His work includes descriptions of native peoples, particularly in reference to the Scythians, inhabitants of the steppes, as well as the Taurians, the inhabitants of the southern mountainous regions. But as does Herodotus, Strabo portrays many of the inhabitants as savages.

In his *Natural History* (*Naturalis Historiae*), Pliny the Elder provides interesting information about the natural environment and rural management in the ancient world. In some instances he makes comparisons between regions, pointing to advantages and disadvantages of the land and climate among the different Greek cities around the Mediterranean and the Black Sea. His brief reports on the Crimean Peninsula appear in Books 4 and 6 of *Natural History*. But despite the book's title, his descriptions focus more on towns and ethnic groups than on natural aspects. Nonetheless, mentions of major land features are essential for understanding aspects of the land as well as the geographic context of accounts in other ancient sources.

Of lesser importance, but with interesting references to Crimea's geography, are the *World Survey* (a loose translation of its Greek name: *Ges Periodos*) by Hecateus of Miletus, the *Enquiry of Plants* by Theophrastus, the *Geographies* by Claudius Ptolemy, and the *Periplus of the Black Sea* by Flavius Arrianus. Hecateus of Miletus also produced descriptions of regions in the Black Sea and a map, both of which were consulted and cited by Herodotus (4, 36).

Although southern Crimea was under the influence of the Byzantine Empire for most of the Middle Ages, no detailed Byzantine accounts of its natural history and geography exist for that time. Towards the end of the Middle Ages, various territories in Crimea were held by other local and foreign powers, particularly Kievan Rus, the Kingdom of Trabezond, the Venetian and Genoese merchants, and finally the Tatar-Mongols.

Kievan Russian travelers visited Crimea often and established an enclave of what today is Kerch. The two geographers who are known to have written about Crimea were Geb Sviatoslav (ninth century) and Afanasiy Nikitin (Yena, Yena and Yena 2007). The former carried out a topographic survey of the Taman and Kerch Peninsulas. Afanasiy Nikitin traveled through Crimea via Balaklava and Caffa (modern Feodosia) on his return trip from India but, as with most other medieval travelers, geographers and writers, there is little mention of aspects relating to the geology, flora and fauna.

Having expelled their Venetian competitors, the Geonoese maintained strongholds on the southern coast of Crimea, in places such as Chersonesos, Chembalo (Balaklava), Soldaia (Sudak) and Caffa (Feodosia). The sources at this time often make reference to the timber resources of the southern mountainous regions (Yena, Yena and Yena 2007). The takeover of the peninsula by the Ottomans at the end of the fifteenth century meant the end of Genoese trade in the region, and somehow a break with Western trade. But in the subsequent centuries, still under Tatar and Ottoman rule and suzerainty, Western travelers visited Crimea. The Italian Emidio d'Ascoli, who lived in Crimea between 1624 and 1634, reports on many natural resources of Crimea, of which the export of timber is the most significant (Yena, Yena and Yena 2007).

The best description of the Crimean Peninsula during the Tatar-Ottoman period was provided by the Turkish traveler Evliya Çelebi. The accounts of his two travels through Crimea in 1641–2 and 1663 were published in his *Book of Travels* (*Seyahatname* in Turkish), which appeared in English translation (Çelebi and von Hammer-Purgstall) in 1834. The section concerning travels in the Crimean Peninsula was published in Russian (Çelebi 1996). Evliya Çelebi provided general descriptions of plants, animals, landscapes, agriculture, ethnic groups and languages. Although it is not a systematic description, as is the case of later travelers, Çelebi's portrayal of the Crimean Peninsula emphasizes the differences in the natural and cultural landscapes and climates of the peninsula and the meaning of these differences for agricultural development in the early seventeenth century.

Before the official Russian annexation of Crimea in 1783, Russian diplomats produced some descriptions of the Crimean landscape. Ivan Sudakov was one of the most notable geographers to produce descriptions (particularly on climatic aspects) of the peninsula in the mid-1500s (Yena, Yena and Yena 2007). Closer to the time of annexation, however, a series of descriptions was available during a brief occupation of Crimea by the Russian army in 1774. Accounts of vegetation and, in general, the geography of the peninsula, were collected and written up by Mikhail Tarkhimovskiy, a future army doctor and contemporary of the notable Russian scientist Mikhail Lomonosov.

The era of naturalists

A new era of interest in Crimea's environmental history begins with the annexation of Crimea by Russia in 1783. Scientific interest in Crimea's natural history developed as part of the Empire's wider interest in the natural resources of the newly acquired territory. Consequently, most of the first naturalists who came to Crimea were commissioned by institutions such as the Imperial Academy of Science of St Petersburg, or directly by the government. Among the most salient of these figures were Vasiliy F. Zuyev, Karl Ludwig Hablizl, Peter Simon Pallas, Frederick K. Biebershtein and Christian C. Steven, all of whom contributed greatly to knowledge of plant and animal species, as well as minerals and rocks, and to understanding the potential for a number of economic activities. But even before Crimea was officially part of the Russian Empire Vasiliy Fyodorovich Zuyev traveled through the peninsula in 1782 collecting and describing fauna, flora and landscapes. His work never made it into a compendium but it was published in a type of almanac in 1783, which for many years remained a guide for naturalists visiting Crimea (Yena, Yena and Yena 2007).

The most solid contributions to the knowledge of Crimea's environment at the end of the eighteenth century were those of Karl Ludwig Hablitz and Peter Simon Pallas. Both naturalists were German émigrés at the service of the Russian Empire. They had a similar academic background and worked together on many projects (Wendland 1992). Their works are still cited by botanists, zoologists, geographers, geologists, historians and archaeologists.

Karl Ludwig von Hablitz (known also as Karl Ludwig Hablizl, sometimes Habliz, or by his Russianized name Karl Ivanovich Gablits) was the first Vice-Governor of Tauria (Crimea). He was the first naturalist to create a solid body of knowledge about Crimean plant and animal species classified according to the Linnean system. His work, published in the *Physical Description of Taurida in Relation to the Three Kingdoms of Nature* (in Russian in 1785), had a great impact on the natural science community and on forthcoming naturalists in Crimea. Peter S. Pallas made various references in his book to the findings and descriptions previously published by Karl Hablitz.

Peter Simon Pallas was born in Berlin in 1741 and received his education at the universities of Halle and Gottingen (Wendland 1992). He received his doctorate in 1760 at the University of Leiden. He was recommended to Empress Catherine the Great and invited to participate in the study of Russia's natural history. In 1767, at the age of 26, he moved to Saint Petersburg where he became a member of the Academy of Sciences (Yena, Yena and Yena 2007). During his early years as a Russian academic, he led expeditions to the Caspian region, the Baikal, and numerous other regions of

Siberia. His publications included memoirs and descriptions of his expeditions, and a volume on the flora of Russia (Wendland 1992).

In 1793–4 Pallas led his first expedition to Crimea. A short report appeared in 1795, entitled *Short Physical and Topographical Description of the Taurian Region*. That same year Pallas moved back to Crimea, where he stayed until 1810. He lived on his estate in Salgirka, which is now part of the Salgirka Park, where the Botanical Gardens of Simferopol are located. During this time he traveled, collected material and continued studying and publishing more writings on Crimea and other regions of Russia. A more detailed compendium of his work, *Travels through the Southern Provinces of the Russian Empire in the Years 1793 and 1794*, was published in 1795 in Russian, and in 1812 in English.

Pallas identified and recorded a number of new plant species, adding information on their ecological context and geographic distribution. In his honor, some species were renamed after him, the best known being *Pinus pallasiana* (Crimean pine). His contribution to Crimea's natural history goes beyond the life sciences; it includes the first account of the geology and landforms of Crimea and his writings also include aspects of ethnography, archaeology and history.

In addition to the naturalists commissioned by the Russian government, a number of independent travelers explored, reported and sometimes collected specimens. Pavyel I. Sumarokov traveled through Crimea in 1799, recording many aspects of its natural history, environment, resources and culture, most notably the medicinal properties of Lake Saki's mud, as well as the advantages of Crimea's mild climate for the treatment of a number of illnesses (Sumarokov 1800, 1803–5). Other travelers who made contributions to our knowledge of the natural history and environment of Crimea include Ivan Murav'yov-Apostolov (traveled in 1820), Frédéric Dubois de Montpéreux (traveled in the 1830s) and Anatole M. Demidov (traveled 1840–2).

Frédéric Dubois de Montpéreux is perhaps the most prominent of the independent travelers. He collected abundant information on ethnography, history, archaeology and natural history, all of which were published in 1843 in his book *Voyage autour du Caucase, chez les tcherkesses et les abkhases, en Colchide, en Géorgie, en Arménie et en Crimée*. Because of the recording and sketches of Crimea's antiquities this book has been regarded as an essential record for further archaeological research. For this and for its description of natural history and environment, this book is perhaps one of those on a par with Peter Pallas's 1812 work.

The naturalists who visited Crimea during the decades following the annexation were a product of the post-Linnaean and pre-Darwinian era. Their main objective was to collect and classify specimens (Farber 2000). This current of scientific thought comprises numerous scholarly figures around the world at the time: Alexander von Humboldt, Sir Joseph Banks, and Meriwether Lewis and William Clark, among others. In fact, a parallel can be traced between the need that the Russian Empire had for the resources of the acquired lands and the need that Thomas Jefferson's United States had for the resources in the newly acquired territories of the Louisiana Purchase and Oregon. Similarly, some of these naturalists also engaged in aspects beyond specimen collection and surveying, including native culture and archaeology.

Other foreign travelers who in the first half of the nineteenth century also contributed to a knowledge of Crimea's natural history, antiquities, and human–environmental relations were Marie Guthrie, Charles Koch and Henry D. Seymour. Many of these travelers also reported on strategic aspects of the peninsula. The most famous examples of this kind of work are Charles Koch's *The Crimea* (1855) and Henry D. Seymour's *Russia on the Black Sea and Sea of Azof: A Narrative of Travels in the Crimea and Bordering Provinces* (1855). Both authors report various aspects of hydrology, climate, topography and overall aspects that at the time would have been of strategic importance, particularly linked to the army and Russian fleet at Sevastopol and Kerch. Interestingly, both books were published in England during the Crimean War.

A big boost to the study of local vegetation in Crimea came in 1812 with the creation of the Tavrida Public Gardens, later named the Imperial Botanical Gardens, and now the Nikitsky Botanical Gardens. The initiative was implemented by the renowned botanist Frederick K. Bieberstein, who was very active in the study of Crimean flora during the first decade of the nineteenth century. The gardens were established in their present location near the village of Nikitsky, six kilometers west of Yalta, with an imperial decree by Tsar Alexander I, and an annual stipend of 1,000 rubles (Bumbyer et al. 1914). Its first director was the renowned botanist, Christian Steven, who directed the gardens from 1812 to 1826.

Originally, the idea of the botanical garden was to develop a center for acclimatization of foreign plants, particularly those from the tropics and subtropics, some of which would eventually make it to other parts of the Empire (Molchanov and Rubtsov 1986). This plan was based on the fact that its location had one of the warmest and mildest climates of the Russian Empire. However, a great deal of research in the botanical garden was also devoted to studying the local vegetation.

Christian C. Steven is also known for his pioneering interest in the endemic flora of Crimea, which is one of the hot topics in Crimea's biogeography today (Yena, Yena and Yena 2007). His successor, Nikolay A. Gartvis, who directed the gardens from 1826 to 1860, focused the institute's research more on subtropical and Mediterranean cultivated plants, particularly for the improvement of grapevine cultivation and the acclimatization of olive, tea and tobacco, and the introduction of decorative plants. It was during this time that the institute for wine production was established in Magarach, near the garden grounds.

Throughout the course of the nineteenth century, other scientists affiliated with the botanical gardens contributed to the study of the local flora, particularly in terms of taxonomic classifications. Research in the Caucasus and Central Asia, some of which involved researchers from the botanical gardens, opened up new ideas concerning the similarities and differences between Crimean and Caucasian floras, thus laying the basis for studies on plant geography.

The era of modern academics

The study of natural history and the environment underwent significant changes around the middle of the nineteenth century as new scientific paradigms appeared. In the broader scientific context, the study of flora and fauna turned more from essentially collecting to universal questions regarding the origins, evolution and interactions among organisms, a global trend stimulated by Darwinian ideas (Farber 2000).

Ideas related to the origins of Crimean flora and the occurrence of endemic flora began to develop rapidly during this period. Furthermore, aspects of ecology began to appear and, most importantly, many studies began to include humans as a factor in the development of natural history. Unlike the previous era, during which most scientists were naturalists researching an array of subjects (flora, fauna, geology, etc.), the second half of the nineteenth century saw the rise of scientists specializing in particular fields.

At the same time as these changes were taking place in the scientific arena, a series of other influences of a social and political nature also changed the ways in which science was conducted in Crimea. The siege and fall of Sevastopol during the Crimean War taught Russian strategists a lesson: a railway was needed to connect the naval bases with the main supply routes. The Russian government thus embarked on a project to connect those lands on the margins of the Empire, particularly those close to the Ottoman Empire, with the centers of power. The opening of the railway was completed to Simferopol in 1876 and to Sevastopol in 1877. Soon other lines followed. These railway lines brought an unprecedented development of agriculture and other industries in Crimea, which required a better knowledge of the landscape and resources (Yena, Yena and Yena 2007). Thus, in addition to studies of the flora and fauna, studies of the geology, soils, hydrology and climatology multiplied during the second half of the nineteenth century.

In tandem with the philosophical, scientific and academic changes, coordinated scientific and public interest in natural history began to consolidate at the end of the nineteenth century. At the initiative of Aleksandr Kh. Steven (Christian Steven's son) and entomologist S.A. Mokrzhyetskiy, among others, the Natural Historical Museum (today's Central Tavrida Museum) was founded in Simferopol in 1899. Mokrzhyetskiy was also among the founders of the Crimea Society of Naturalists and Nature Lovers. These events not only fostered more research: they also popularized the natural riches of the peninsula and, by the beginning of the twentieth century, Crimea had evolved into a natural attraction for educated tourists. This can be seen in *Crimea: A Travel Guide*, edited by K. Yu. Bumbyer et al. (1914) which, although a travel guide, contains scientific articles on flora, fauna, climate, ecology and many aspects of Crimean geography, written by prominent scientists of the time—such as N.I. Andrusov, S. A. Zyernov, N. N. Klyepyenin, S. A. Mokrzhyetskiy and S. A. Yanata. Despite being a century old, this book is of such scientific and historical importance that it is still cited in scientific papers. In particular, it portrays the state of natural history at the time, which can be used as a baseline for evaluating environmental change in Crimea.

During the nineteenth century, very few practitioners of natural science researching Crimea were based in local institutions. The majority of scholars involved in the study of natural history in Crimea were based in St Petersburg, Moscow, Kiev and Odessa. Not until the first two decades of the twentieth century did the local institutions become consolidated into a body of local scientists. Central to this change was the creation in 1918 of Simferopol University, the precursor of the current V. I. Vernadskiy Tavrida National University. In subsequent years, filial institutes of the Academy of Sciences of the USSR, as well as a number of pedagogical institutes, appeared.

The study of flora

The amount of information amassed by naturalists and travelers from the 1780s to the 1850s was valuable for later studies. The importance of their work is demonstrated by the tremendous increase in the number of recognized plant species between 1785 and 1860. Thus, Hablitz's work in 1785 reports 521 species, Pallas's work at the turn of the century reports 990 species, and *Flora Rossica*, published in 1853, reports 1,458 species for Crimea (Yanata 1914). During the second half of the nineteenth century, many academics made an impact on the study of flora. Worthy of mention is the work by Vladimir N. Aggyeyenko, who published extensive lists of flora of the peninsula, which culminated in the publication of *Flora Taurica* (1890), a volume that listed 90 percent of the then known flora of Crimea. The current number of species is 2,536 (Yena 2012).

As for the study of vegetation in the first half of the twentieth century, a number of scientists, some of them locally based at Tavrida National University or the Nikitsky Botanical Gardens, had an impact, which in the new century turned more to evolutionary and ecological aspects. One of the best known worldwide is Yevgeniy V. Wulff (or Vul'f, in Russian transliteration), who wrote numerous articles on a variety of botanical and ecological topics. In addition to his numerous contributions to new identified species to *Flora Kryma* (also known as *Flora Taurica*), Wulff published a number of articles on the origins of Crimean vegetation, the yaila vegetation, the naturalization of subtropical wild and cultivated plants, and a number of other subjects that gave the Nikitsky Botanical Gardens a reputation in botanical and ecological studies.

Other prominent botanists before 1950 include Vatslav Stankevich, Polish in origin, who devoted a great deal of research to the study of forests in Crimea and the Caucausus, and Sergey S. Stanko, who took charge of Wulff's work *Flora Kryma* (Yena, Yena and Yena 2007). Worthy of mention during this period is Vladimir P. Malyeyev, who wrote extensively on the origins of Crimean flora, and Nikolay A. Troitskiy, who in the late 1940s conducted a series of geobotanical surveys of the steppe vegetation communities in the Crimean plains, which today have disappeared under the plough.

One circumstantial aspect of the study of Crimean vegetation and soils concerns the research carried out and published by the German geobotanist Heinrich Walter, who was appointed chief of the Nikitsky Botanical Gardens during the Nazi occupation of the peninsula in 1943 (Walter 1980; Molchanov and Rubtsov 1986). In his memoirs he recounts passages of field research under military protection due to partisan activities, particularly in the mountains (Walter 1980). His Crimean research was published in *Die Krim: Klima, Vegetation und landwirtschaftliche Erschliessung* (1943), which contains photographs and sketches of vegetation. Unfortunately, his work is surrounded by the negative aura of the brutal Nazi occupation of the peninsula. Despite this fact, Heinrich Walter was one of the first to promote widely the different aspects of Crimea among botanists, ecologists and biogeographers outside the former Soviet Union in the second half of the twentieth century, particularly through his book *Die Vegetation Osteuropas, Nord- und Zentralasiens* (1974), as well as in numerous other geobotanical treatises.

During the first two decades of the second half of the twentieth century, several botanists made significant contributions to our knowledge of Crimean flora. These include Nikolay I. Rubtsov, whose various publications on the flora of Crimea have been widely cited, as well as those by M. S. Shalyt,

V. G. Mishnyov, I. V. Golubyeva, V. N. Golubyev, Ye. F. Molchanov, M. A. Kochkin and I. V. Kryukova. Among those involved in the study of vegetation, strictly understood as the study of communities, are several locally based botanists, such as N. N. Dzyens-Litovskaya, who provided a tremendous amount of knowledge on steppe vegetation and soils; Vladislav V. Korzhyenyevskiy, whose work in the area of flora and vegetation is of great prominence; Tatiana Larina, who devoted research to the Sub-Mediterranean vegetation, and A. V. Yena (Crimea Agrarian University), who has dedicated research to the thorough study of endemic flora. In the most recent decades, most of the local scientists working on various aspects of botany, plant ecology and biogeography are associated with the Nikitsky Botanical Gardens, Tavrida National University and Crimea Agrarian University. A number of scientists based in institutions in mainland Ukraine have also contributed to the study of Crimean flora, among who the most influential are Yuriy R. Shelyag-Sosonko and Yakov P. Didukh.

The study of fauna

Previous to the era of academics, substantial work was done in the field of zoology, particularly in identification of species. Karl L. Hablizl's *Physical Description* (1789) includes a list of 9 mammals, 38 birds, 12 freshwater fishes, 17 sea fishes, 11 mollusks and crustaceans, and 9 reptiles and amphibians. Pallas's work does not contribute much to the zoology, since he was more interested in flora, but numerous mentions of animals and hunting practices are of great value.

Throughout the first half of the nineteenth century eminent zoologists contributed to the knowledge of Crimean fauna. In the first half of the century, the most notable scientist was A. Syevastyanov, who contributed to the study of Crimean mammals, birds, fishes, amphibians and reptiles (Dulitskiy 2001). In the second half of the century zoological studies increased, this time focusing more on each of the groups of the animal kingdom. The most prominent zoologists of this period were A. A. Brauner, L. Kh. Ibri, Kh. Goebel and S. A. Mokrzhyetskiy and A. M. Nikolskiy (Dulitskiy 2001; Yena, Yena and Yena 2007).

Throughout the rest of the twentieth century many scientists based in local and other institutions of the Soviet Union contributed to our knowledge of Crimean fauna—K. A. Yefyetov, L. G. Apostolov, V. N. Popov, S. P. Ivanov, N. N. Scherbak, Yu. V. Kostin and A. I. Dulitskiy, among others. Worthy of mention here are the treatises by N. N. Scherbak, *Amphibians and Reptiles of Crimea* (1966); Yu. V. Kostin, *Birds of Crimea* (1983); and A. I. Dulitskiy, *Mammals: History, Status, Conservation and Perspectives* (2001), which are widely used and referenced.

Earth sciences

Rocks and minerals, as well as all other aspects of geology, have been subjects of interest since the time of the early naturalists. Interest in geology developed rapidly throughout the nineteenth century, so that by the end of the century a great number of published works and geological maps existed. At the turn of the century interest in the Quaternary—an important topic for environmental history—began to develop. One of the outstanding figures who did pioneering work in Tertiary and Quaternary geology was Nikolay I. Andrusov, who should be considered the father of Crimea's Quaternary geology. His work on defining the chronology of marine and alluvial terraces, and his writings on Crimea's Pleistocene landscapes, are of major importance for future generations of scientists in the field. The publication of his work on alluvial and marine terraces (e.g., Andrusov 1912) opened up a new era in research, particularly in regard to sea-level changes in the glacial phases of the Pleistocene.

The first decade of the twentieth century saw a rapid increase in papers and treatises on the peninsula's fossil record and its geological history in the context of mountain formation in Europe and Asia. In parallel with this development, marine studies began to increase dramatically. For this period, the contributions of Vladimir I. Vyernadskiy to the geology and physical geography of Crimea and the foundation of research stations are worthy of mention, as is the work of Nikolay A. Golovkinskiy in the areas of stratigraphy, hydrogeology, sedimentology and geomorphology.

Among geoscientists in Crimea in the first half of the twentieth century, it is important to mention Aleksey A. Borisyak, who is known for his study of the vertebrate fauna of the Tertiary period in Crimea, a topic of great relevance in the evolution of camelids and equids worldwide. In the area of magmatic volcanism an important study was carried out by Alyeksandr F. Sludskiy, who made an extensive study of the Middle Jurassic volcanic massif of Kara-Dag (now a nature reserve) and established a research station there. The study of karstic features, which are so abundant in the peninsula, received attention from many geologists, primarily A. A. Krubner. Marine geology also became a significant geoscience field, headed by, amongst others, Igor V. Kurchatov.

During the second half of the twentieth century, research in the earth sciences in Crimea enjoyed an unprecedented boom. Local geologists, as well as those from Moscow, St Petersburg (then Leningrad) and Kiev, made significant contributions to our understanding of the many geological and geomorphological particularities of the peninsula. The most notable figures in the area of Crimean geology in the middle third of the twentieth century were I. S. Blagovolin, V. T. Lyebyedinskiy and Ya. D. Kozin. Mikhail V. Muratov, based in Moscow, is well known for his life-long study of Crimean geology and for his widely cited publications that summarize the geological history of the peninsula (see Muratov 1974).

During this time many geological schools from around the USSR took students to Crimea, where a great number of textbook examples were to be found throughout the peninsula, particularly in the mountains. Moscow State University (abbreviated MGU in Russian) began their scholarly field trips in 1935. An MGU-sponsored geological camp station was established in 1957 near Bakhshisaray with the purpose of supporting groups of students during summer fieldwork (Milanovskiy 1997). The station is now known as the A. A. Bogdanov Crimean Scientific Center of the Geological Faculty of MGU.

Among the representatives of the various subdivisions of the earth sciences in Crimea in the second half of the twentieth century are:

karst studies: N. A. Gvozdetskiy, V. N. Ivanonv, V. N. Dublyanskiy and V. A. Vakrhushev;
mud volcanism: E. F. Shniukov;
hydrology: A. N. Olifyerov and R. A. Flilyenko;
geomorphology: A. A. Klyukin;
paleontology: N. I. Lysenko;
geophysics: I. I. Popov;
mineralogy: V. A. Abruchyev, Yu. A. Polikuov, V. S. Tarasenko and A. I. Thishchenko.

The study of soils is one area that deserves special attention in the geological, biological and geographic disciplines. Although numerous works addressed the topic in the eighteenth century, the work of Vasiliy V. Dokuchayev, the father of Russian soil science, made a particular impact in the study of the soils of Crimea. He identified a type of Crimean chernozem different from those in the mainland. Inokentiy P. Gerasimov, a renowned physical geographer from MGU, made efforts to include the typical soils of Crimea, such as the Crimean southern chernozem and the cinnamonic soils, into the classification of soils of the USSR (Gerasimov 1954).

Many pedological studies were carried out in Crimea during the twentieth century by renowned scientists, including N. N. Klepinin at the beginning of the century. During the second half of the century, work by Nina N. Dzyens-Litovskaya had great impact not only in pedology but also in geobotanical studies, particularly with regard to the steppe region of the peninsula (e.g., Dzyens-Litovskaya 1970). With regard to mountain soils, Mikhail A. Kochkin (1967), and more recently Natalia A. Dragan (2004), have worked on a more detailed classification of Crimean soils, reporting about 400 soil types (Yena, Yena and Yena 2007).

Interest in the climate of Crimea dates back to naturalists in the late eighteenth century but systematic measurement and study did not begin until the end of the nineteenth century. The fields of meteorology and climatology developed rapidly in the twentieth century in tandem with other areas of research, particularly botany and agriculture and, later, with air navigation (Vyed' 2000). In fact, it was at the Nikitsky Botanical Gardens that the field of climatology began to take place in Crimea (Vyed' 2000). Studies in this area have increased as the network of meteorological stations in the peninsula has grown from a handful in the main cities at the beginning of the 1920s to, currently, 25 meteorological observation stations. About the same number of meteorological stations form part of the hydrological marine, agricultural and air navigation observation stations (Vyed' 2000). Two scientists in the field of climatology during the first half of the twentieth century were P. A. Kondrat'ev, who published the first climatic atlas of the peninsula, and geographer A. A. Borisov, who also wrote about climate change in the geological past and was the first person to compare systematically the climate of southern Crimea with that of the Mediterranean. Other scientists in the mid- and late twentieth century were I. I. Babkov, D. A. Burtsyev and V. I. Bazhov (Yena, Yena and Yena 2007). More recently climatology is represented by such scholars as Avgust N. Olifyerov, Ivan. P. Vyed' and Yelyena I. Yergina, all at Tavrida National University.

Geoarchaeology and paleoecology

Given the themes of this book, it is important to mention researchers who have contributed to reconstructing past environments, particularly in the fields of geoarchaeology and Quaternary paleoenvironments. Of note here are studies by Lavrentiy A. Moysyeyev and Stanislav F. Strzhyelyetskiy on the ancient Greek territory of Chersonesos, by L'yev V. Firsov (based in Novosibirsk), who introduced the radiocarbon technique to many studies dealing with ancient settlements and sea-level change, and Aleksander A. Klyukin, who has contributed to our understanding of prehistoric and historic landscape change. Recent work in the area of geoarchaeology includes research in the Paleolithic sites of the Outer and Inner Ridge (Chabai, Marks and Monigal 1999, 2000) and the Heraklean Peninsula (Cordova and Lehman 2005 and Cordova et al. 2011).

Pioneering work on pollen research by Artyushyenko and Mishnyov (1978), aimed at reconstructing yaila vegetation, has recently been followed up by a study of pollen from sediments in the Yaltinskaya Yaila (Cordova et al. 2011 and Chapter 6 in this book). Studies of macro-botanical remains of the Pleistocene were carried out by Gubonina (1985) at a number of archaeological sites in the piedmont. Later, Natalia P. Gerasimenko (2004,

2007) studied pollen from deposits associated with Pleistocene archaeological sites in the Outer and Inner Ridges. Studies of pollen focused on the Holocene vegetation changes in the southwest part of the Peninsula have addressed many of the old questions about the origin of Crimea's vegetation and the impact of humans (Cordova and Lehman 2003, 2005). Work by paleobotanists such as Zoya Yanushevich (Yanuchevitch) from Odessa and Galina Pashkevich from Kiev has also linked vegetation to the ancient groups who inhabited Crimea, particularly through domesticated plants, a topic discussed in Chapters 6 and 7.

Geography, landscape ecology and conservation

During the second half of the twentieth century, interest in conservation ecology developed among local institutions as environmental awareness grew in response to the rapid transformation of the land that led to a loss of habitats and threats to biodiversity and rare species. Interestingly, the field of ecology became part of the discipline of geography, particularly physical geography, geoecology and landscape ecology (or its Russian equivalent, *Landshaftovedeniye*), all of which are here considered as landscape studies.

The physical geography tradition can be seen in a number of publications in the middle part of the twentieth century, notably the work of Russian geographer Boris F. Dobrynin (based at MGU), who included a chapter on Crimea in his book *Physical Geography of the USSR: The European Part and the Caucasus* (1948). Subsequently, a strong physical-geography tradition began to develop on Crimean soil, focusing on Crimea itself, as part of a Krymovyedeniye (Crimean scholar interest), particularly in Tavrida State University. At least three schools of thought can be distinguished in Crimean landscape studies in the last four decades of the twentieth century: the deterministic school, the regional uniqueness school, and the statistical-probabilistic school.

The deterministic school focused on regularities and typologies, a view focused on the big picture of nature, a systematic whole of organized components. This school became popular in the late 1960s and its key exponents included leading geographers such as Gregory Ye. Grishankov, who created a detailed typological map of Crimean landscapes, Mikhail A. Kochkin author of the key work *Soils, Woodlands and Climate of the Mountainous Crimea and the Ways of their Rational Use* (1967), and a number of botanists and geobotanists, which included such figures as V. N. Golubyev, T. D. Vodop'yanova, and N. I. Rubtsov.

Around the same time, the regional uniqueness school of thought appeared as an antithesis to the deterministic school, stressing the idea of regionalism and emphasizing the uniqueness of Crimean landscapes. This school promoted conservationism, and it is the precursor of many studies of nature reserves and parks today. Representatives of this school included Vladimir G. Yena and Pyotr D. Podgorodyetskiy, who wrote extensively on Crimea.

Finally in the 1980s, an alternative statistical-probabilistic school of thought began to gain momentum, one influenced by quantitative ideas, with a strong emphasis on collecting experimental data and seeing nature in terms of interacting factors. Most proponents of this school were geographers in the areas of climatology, geomorphology and geobotany, including Aleksandr A. Klyukin, Vladimir A. Bokov and Vladislav V. Korzhenevskiy.

In the twenty-first century, the ideas of these schools have faded somewhat and merged, because many of the students and disciples have combined the three approaches, particularly in the context of globalization and new issues facing Crimea—particularly rapid development, privatization and growing tourism. More recent volumes, containing geographic and landscape information as well as new ideas drawing on the aforementioned schools of thought, include *The Geography of Crimea* (Bagrova, Bokov and Bagrov 2001) and the edited volumes *Modern Landscapes of Crimea* and *Adjacent Seas* (Pozachyenyuk 2009a).

Recent developments

The collapse of the Soviet Union in 1991 and the ensuing independence of Ukraine, to which Crimea belonged, brought some structural changes to the scientific organizations. Many of the institutions and branches came to depend on the Ukrainian Academy of Sciences and, above all, on decisions in Kiev. However, a number of scientists from other countries of the former Soviet Union, particularly Russia, are still involved in many research projects in Crimea.

Today dozens of large and small centers of research include universities, pedagogical institutes, research institutes, and research stations located in numerous areas of the peninsula. The leading institutions of local research based in Simferopol include V. I. Verdadskiy Tavrida National University; the Crimean Agrotechnological University; the Crimean Academy of Sciences; the Crimean Scientific Center of the National Ukrainian Academy of Sciences; the Crimean Office of Geodetics, Cartography and Geoinformatics; the National Academy of Nature Conservation and Resort Development; and the Crimean Center of Hydrometeorology. Institutes and centers based in Sevastopol include the A. O. Kovalevskiy Institute of Biology of the Southern Seas, the Marine Hydro-Geophysical Research Institute of the National Ukrainian Academy of Sciences; and the Black Sea Division of the Marine Hydrophysical Institute; whilst the National Institute of Marine Fisheries and the Oceanography Institute are based in Kerch. In Yalta, in addition to the

National Research Center at the Nikitsky Botanical Garden, the Magarach National Research Institute for Viticulture and the National Institute of Research for Climatological Medicine and Therapeutical Methods, are centers of high academic reputation. Many other institutes and research stations exist in nature reserves throughout the peninsula.

Although research slowed down due to the political and economic changes in the 1990s, studies in the area of nature and ecology took off again in the first decade of the twenty-first century. The post-Soviet opening up meant more contact with foreign research institutions and more cooperation between local, national and foreign scientists. As a result, Crimea's questions regarding Crimea's biogeographic aspects—such as endemic and relict species, as well as climate and vegetation changes—are now treated more as European, Eurasian and global issues rather than as local phenomena. Most importantly, more and more publications on Crimea's nature, environmental history and conservation are now appearing in international journals.

CHAPTER 3

THE PHYSICAL ENVIRONMENT

Physiography and landforms

The Crimean Peninsula comprises three main physiographic units: the plains, the piedmont, and the mountains (Figure 1.1). The plains occupy two-thirds of the peninsula, and their topography varies from fairly flat in the center to rolling plains and low hills in the Tarkhankut and Kerch peninsulas. The piedmont comprises the Outer and Inner Ridges (Figures 3.1 and 3.2A–B), both of which have the shape of a *cuesta*, a term that refers to a ridge with a gentle slope on one side and a steep step on the other (Figure 3.3). The Outer Ridge forms a slight ascent from the plains to a maximum elevation of 344 m, dropping into a valley from where the Inner Ridge ascends, reaching a maximum elevation of 564 m. The Main Ridge comprises the mountains proper, which also have an asymmetrical structure with a much steeper slope on the south side and which are capped by a series of plateaus known as yailas, or collectively as the Yaila (Figure 3.4).

The yailas form two groups, the western and eastern yailas, separated by the Angara Pass (Figure 3.5). The western yailas (Baydarskaya, Ay-Petrinskaya, Yaltinskaya, Nikitskaya, Gurzufskaya, Babugan and Chatyr-Dag) constitute the highest parts of the mountains, with elevations ranging mainly between 700 m and 1,400 m. The highest point is located at Roman Kosh (1,545 m) on the southern edge side of the Babugan Yaila. The eastern yailas (Demerdzhi, Dolgorukovskaya and Karabi) are lower, usually with maximum elevations ranging between 700 m and 1,250 m. The highest point is 1,259 m, on the south side of the Karabi Yaila. The topography of the yailas is characterized by rolling hills and shallow depressions in calcareous rocks.

Tectonics and geology

The Crimean Peninsula occupies portions of two tectonic units: the Scythian Plate and the Alpine Tectonic System of the Crimean Mountains (ATSCM)

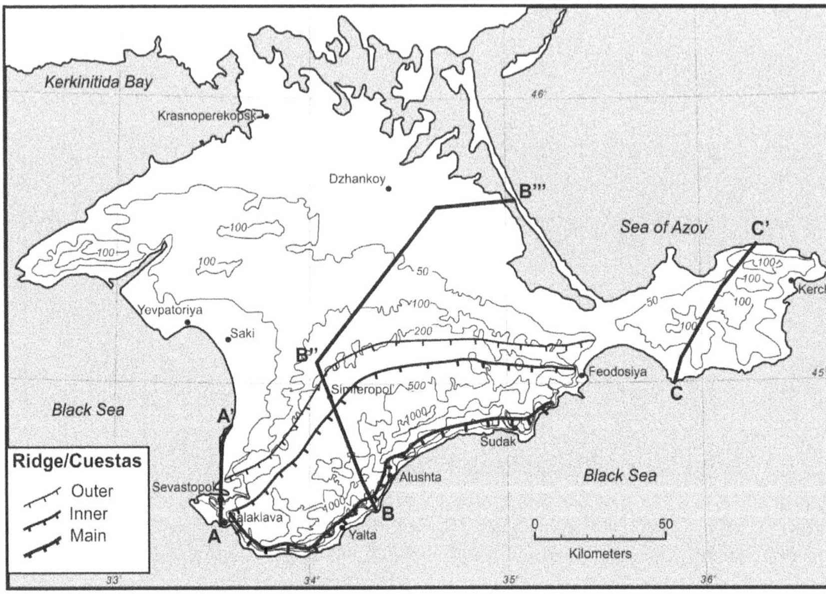

Figure 3.1: Topography, the ridges, and location of the geological transects depicted in Figure 3.2. Source: Cordova et al. (2011).

(Figure 3.6). In the broader regional context, these tectonic units are a deformed edge of the Eastern European Platform, with which they share a Pre-Cambrian basement (Muratov 1974; Permyakov and Maidanovich 1984; Podgorodyetskiy 1988). The boundaries between the Eastern European Platform proper, the Scythian Plate, and the ATSCM are marked by deep faults, marking important changes in the sedimentary history and deformation of the rocks (Pasynkov 2003a).

The Scythian Plate supports a pack of sediments ranging from Paleozoic to Cenozoic. Its northern boundary, marked by a fault running along the Sivash, is considered part of a larger graben structure running east–west from the Danube to the Don Estuary (Permyakov and Maydanovich 1984). The Scythian Plate is further subdivided into several smaller depressions and uplifted blocks (Figure 3.6). In the east, the Indolo-Kuban Depression contains a wide range of marine sediments that in the south (Kerch Peninsula) were deformed by folding (Figure 3.2C). In the west, the Al'ma Basin contains a sequence of Neogene sediments. This basin collects the drainage of the western half of the mountains and directs it into the Gulf of Kalamita. The Tarkhankut anticline constitutes an uplifted block capped by deformed sediments of Neogene age. In the south-central part of the Scythian Plate, the Simferopol Rise is an extension of the monoclinal structures of the ATSCM that has also been important in dividing the drainage of the main rivers

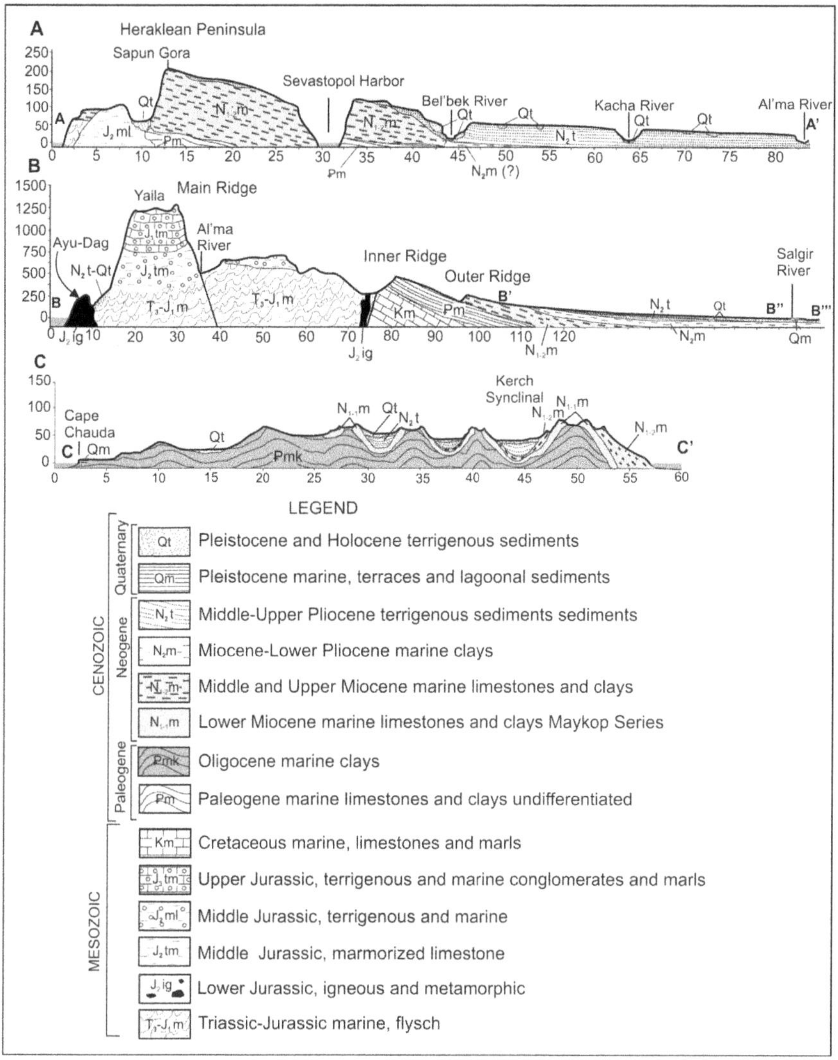

Figure 3.2: Geological cross-sections. Source: Cordova et al. (2011).

between those flowing into the west coast (via the Al'ma Basin) and those flowing to the east and northeast into the southern Sivash (i.e., the Indolo-Kuban Basin). The ATSCM corresponds to the structures forming Main Ridge, the Inner Ridge, the southwestern third of the Kerch Peninsula, and the base of the Heraklean Peninsula. The ATSCM is composed of rocks of Triassic and Jurassic age at its core (central and southern side) and Cretaceous and Tertiary age on its exterior (northern flanks) (Figure 3.2B), most of which are folded, fractured and faulted, forming complex tectonic structures (Figures 3.2 and 3.6).

Figure 3.3: Inner Ridge in the background. The Bel'bek river valley is in the foreground.

Figure 3.4: The southern side of the Chatyr-Dag Yaila. View from the Chistoye village. The valley in the foreground is the Angara Pass.

The surface geology of the Crimean Peninsula comprises a wide range of rocks formed during the Mesozoic and Cenozoic eras (Figure 3.7). The most common rocks in Crimea are of sedimentary origin, of which shale, flysch, clays, limestones, marls, sandstones, and conglomerates are the most common

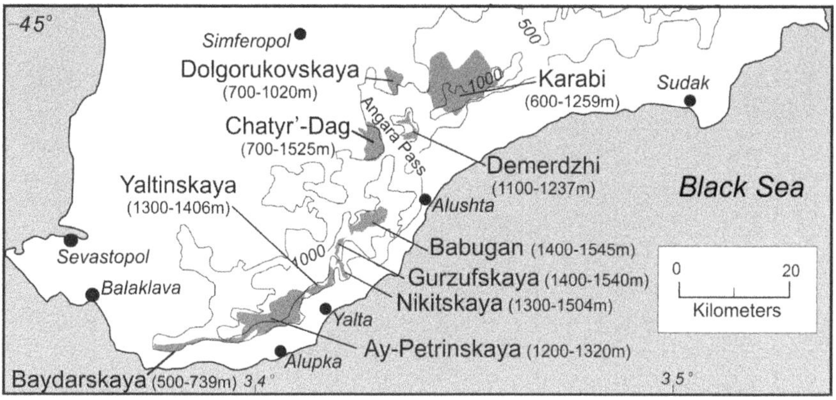

Figure 3.5: The Crimean yailas with their approximated minimum elevations and highest elevations.

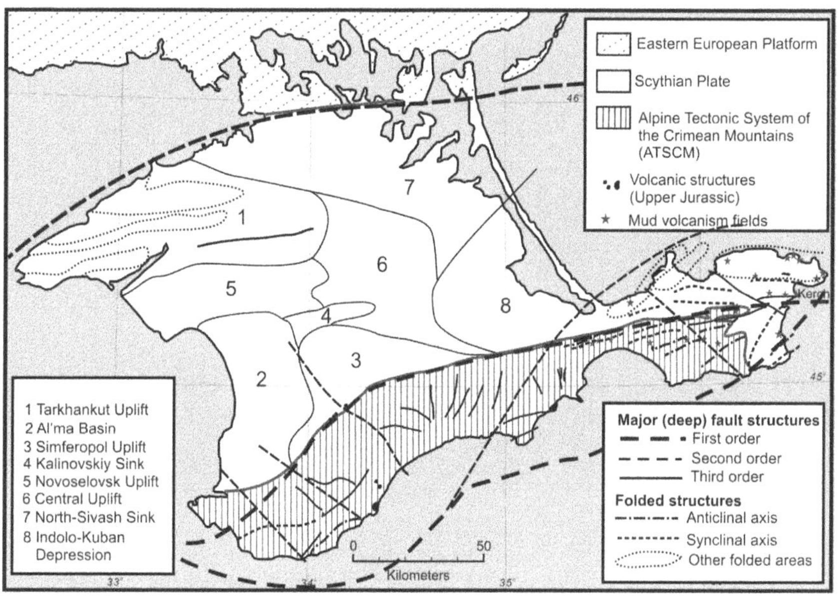

Figure 3.6: Tectonic structures. Source: Pasynkov 2003a.

(Lyebyedinskiy 1982). Unconsolidated sediments, mostly of Pliocene and Pleistocene age, include alluvium, eolian (mainly loess) and marine sediments, and a variety of sediments deposited in karstic depressions in the yailas.

Rocks of volcanic origin in Crimea correspond to the andesitic-basaltic structures exposed along the southern coast and in some localities in the northern slope of the mountains, particularly in the Al'ma river valley. The most notable of these structures exposed by erosion are the Ayu-Dag mountain (Figure 3.2B), and the Kara-Dag structure (near Feodosia), dated to

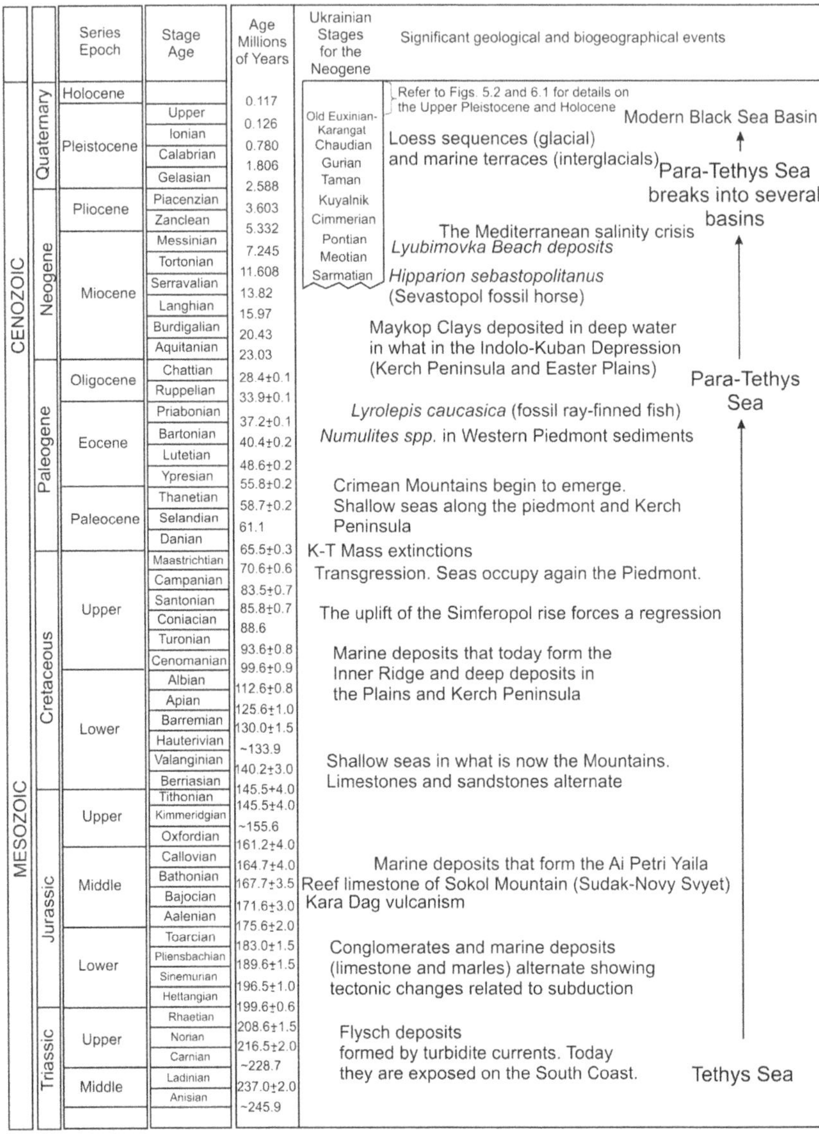

Figure 3.7: Generalized stratigraphic geological chronology and regional chronologies with broad geological events in Crimea. Sources: The International Commission on Stratigraphy (2013), Frolov (1988), Barg and Ivanova (2000), Matoshko, Gozhik and Semenenko (2009), with further additions by the author.

the Middle Jurassic (Bajocian) (Meijers et al. 2010) (Figure 3.7). Other structures include dikes of intrusive (plutonic) rock in the Bodrak river valley, at different points along the south coast at the base of the sequence in Mys Fiolyent, on the south coast of the Heraklean Peninsula (Figure 3.2A), and

Figure 3.8: Mud volcano in the Kerch Peninsula. Professor Aleksandr Klyukin and Paul Lehman in the photograph.

other points along the western part of the south coast (Slavin 1975; Lyebyedinskiy 1982).

Mud volcanoes are interesting geological features found in the Kerch Peninsula (Figure 3.8). Their formation is, nonetheless, not associated with magmatic chambers—as in the case of true volcanism. They are the result of gas pressure building up in underlying rocks, and thus have potential for gas and oil prospection (Kholodov 2002).

Rocks of metamorphic origin are practically absent in Crimea. Clasts of metamorphic rocks of the Precambrian and Paleozoic ages are found in Jurassic conglomerates, suggesting that they were eroded from now-disappeared older mountains perhaps associated with the basement Eastern European Platform (Muratov 1974; Slavin 1975; Permyakov and Maydanovich 1984). But besides this peculiar occurrence, no metamorphic rocks are exposed in the peninsula. The so-called "marble limestone" of Middle Jurassic age represents a partial process of metamorphization, which is not enough to be considered metamorphic (Muratov 1974). This marble limestone can be seen along the south coast, particularly around Balaklava, in the southeastern part of the Heraklean Peninsula, and at numerous areas of the south coast and in the interior valleys (Lyebyedinskiy 1982). In the Sudak-Novy Svyet area, a massive structure of marbled limestone of a former reef deposit forms the Sokol Mountain and the hill upon which the fourteenth century Sudak Fortress was built (Figure 3.9). Marble Cave (*Mramornaya*

Figure 3.9: View of the Sokol Mountain from the Genoese fortress in the Sudak-Novy Svyet area.

Peschera), located on the northern slope of the Chatyr-Dag Yaila, is another typical locality for marble limestone outcrops. The marble limestone has been heavily quarried particularly in the Balaklava area, used for buildings in many Crimean cities and exported to other parts of the former Soviet Union. Among other construction projects, it was used for embellishment of some subway stations in Moscow and Kiev.

Brief geological history

The geological chronology of the Crimean Peninsula (i.e., Muratov 1974) has been widely used as a reference for many decades. But it has recently been challenged and modified (i.e., Frolov 1998; Barg and Ivanova 2000; Pevzner et al. 2003; Matoshko, Gozhik and Semenenko 2009). The modifications include re-definition of some names and boundaries and, more important, a shifting of ages based on recent age measurements (i.e., paleomagnetic dating, potassium-argon), as well as through correlation with the internationally recognized geological chronology (Figure 3.7).

Geological units of Precambrian and Paleozoic age have been reported from boreholes at great depths as (as mentioned above) clasts embedded in conglomerates of younger ages. The oldest visible rocks are the Tauric Group of the Upper Triassic and Lower Jurassic age exposed along the south coast and in the river valleys on the northern slope of the mountains (Muratov 1974; Slavin 1975; Permyakov and Maidanovich 1984). The Tauric Group comprises formations of marine sedimentary rocks, mainly shales, flysch, sandstones, limestones, and conglomerates. The sediments that form the flysch units developed under turbid currents in an underwater tectonic depression associated with a subduction trench (Meijers et al. 2010).

The Middle and Upper Jurassic formations consist of limestones, marls, conglomerates, sandstones, and mudstones (Figure 3.7). The formations of this series lie unconformably on one another, indicating alternating periods of sedimentation and erosion caused by cycles of tectonic movements. Tectonism and volcanism have also shaped the structural geology of the Jurassic, as is the case of the mild metamorphism that formed the marble limestone in the Middle Jurassic (Figure 3.7). Tectonic movements during the Upper Jurassic were characterized by plate subduction resulting in the arc volcanism that produced the Ayu-Dag and Kara-Dag volcanic structures (Meijers et al. 2010) (Figure 3.7). Shallow seas behind this volcanic arc produced the carbonated rocks and reef limestones that form the upper parts of the Main Ridge and essentially most of the rocks exposed in the yailas (Figure 3.2B).

The Lower Cretaceous formations lie almost conformably on the Upper Jurassic formations, suggesting a continuum in deposition or a lack of erosion between the two periods. The Crimean Cretaceous is composed of limestones and marls exposed mainly in the lowest sedimentary sequences of the Inner Ridge (Figure 3.2B). Cretaceous marine rocks have also been found in boreholes in the piedmont and the Kerch Peninsula. The relatively stable depositional environment of the Cretaceous contrasts with the subsequent period, the Paleogene, during which intense deformation of older rocks occurred through uplift and folding produced by the orogenic development responsible for the formation of most mountain systems extending from the Alps to the Tian Shan and Himalayas, which explains the designation Alpine Tectonic System of the Crimean Mountains (ATSCM). During the Paleogene, marine sedimentary deposition was absent in the Main Ridge, suggesting that the mountains have already emerged from the sea by folding and uplifting, and instead they were exposed to the erosional processes that provided some of the clastic material for conglomerates of Paleogene and Neogene ages in the piedmont and plains (Frolov 1998).

In the regional geological context, the marine deposits of the Cretaceous and Paleogene form part of the northwestern portion of the Tethys Ocean (Muratov 1974). By the end of the Oligocene, the Tethys Ocean became partitioned, resulting in several seas. One of these was the Paratethys Sea, which formed a series of shallow basins extending from Austriato Kazakhstan. These basins in turn were connected with other basins of the former Tethys Ocean, such as the Mediterranean Sea and the Indian Ocean (Schulz, Vakarcs and Magyar 2005).

The marine sedimentary sequences of the Paleogene (i.e., the Paleocene, Eocene and Oligocene periods) in Crimea are dominated by limestone, marls and clays, whose formations are exposed in the valleys cutting through the Inner and Outer Ridges, in the depression that forms Sevastopol Bay, and several localities of the central and southern parts of the Kerch Peninsula (Figure 3.6). In the western part of the piedmont, the most prominent of the Paleogene fossil-bearing formations are the Middle Eocene Nummulitic Limestone, characterized by *Nummulites* sp., a coin-like organism, and the Upper Eocene brown marls, which contain fossil ray-finned fish *Lyrolepis caucasica* (Yena and Yena 2010). In the western part of the piedmont and the Kerch Peninsula, marine deposition occurred as a result of the sinking of the Indolo-Kuban Depression (Muratov 1974). In this region, the Maykop clays are one of the thickest and most widespread geological units deposited through the Oligocene and the early Miocene (Figure 3.7).

Although the chronostratigraphic scheme of the Ukrainian and Crimean Neogeneis currently changing, it is still possible to reconcile some of the older geological stratigraphic models (e.g., Muratov 1973; Veklich and Sirenko 1976) with the more recent ones (e.g., Barg and Ivanova 2000; Pevzner et al. 2004; Matoshko, Gozhik and Semenenko 2009) and with those employing international chronostratigraphy (Figure 3.7).

During the Lower Neogene (i.e., the Miocene) marine deposition occurs in most of the plains, the northern and eastern parts of the Kerch Peninsula, the Outer Ridge, and the Heraklean Peninsula (Figure 3.2). During the Upper Neogene (i.e., the Pliocene) the gradual transition from marine to terrestrial deposition begins to take place in the plains. This process is represented in the regional chronostratigraphy as the Sarmatian, Maeotian, and Lower Pontian stages of the Middle and Upper Miocene; and the Upper Pontian and the Cimmerian stages of the Early and Middle Pliocene (Pevzner et al. 2004; Matoshko, Gozhik and Semenenko 2009).

The Sarmatian stage in the piedmont is represented by limestones and marls with members of sand and gravel, suggesting that the shallow sea was subject to alluvial influence from the streams coming from the ancestral Crimean Mountains. These gravel and sand facies layers have yielded a number of vertebrate faunas (Barg and Ivanova 2000). One notable finding is by Aleksey A. Borisyak inside the city of Sevastopol, where a primitive form of horse, *Hipparion sebastopolitanus*, was found in association with extinct species of camels and giraffes (Borisyak 1914). Although it has been widely believed that at this time the ancestral Crimean Mountains were an island, the recurrent presence of vertebrate faunas in many parts of western Crimea suggests connection with the mainland (Pevzner et al. 2004). The Maeotis stage (Middle Miocene) and Pontic stage (Upper Miocene) correspond to the Paratethys Sea, which at the end of the Miocene began to evolve into separate basins—Black Sea, Caspian and Aral (Schulz, Vakarcs and Magyar 2005).

During the Pontic and Cimmerian stages, a more permanent land bridge began to form between the Crimean Mountains and the mainland to the north. The best expression of the sequence of events leading to this development is represented in the sedimentary sequences exposed along the cliffs of the west coast, between Sevastopol Bay and the mouth of the Bulganyak River, particularly in the area of Lyubimovka Beach, just north of the mouth of the Bel'bek River. The original Mio-Pliocene stratigraphy in this area (Vyeklich and Siryenko 1976) has been modified by new magnetic dates and correlations (Barg and Ivanova 2000; Pevzner et al. 2004). Parallel with this chronological adjustment, the interpretation of some depositional environments in the stratigraphy has been modified. Thus, deposits of the early Pliocene, originally interpreted by Vyeklich and Siryenko (1976) as fully continental, have now been identified as lagoonal and deltaic with frequent marine transgressions. In more recent studies, correlations with the Mediterranean have revealed that during the middle Pontian (Upper Miocene) significant marine regressions occurred, probably associated with the drying of the Mediterranean during the Messinian stage (Matoshko, Gozhik and Semenenko 2010).

In the global context, the Pliocene has received attention for a warm phase during the Piacenzian stage (Figure 3.7). Vegetation and soils dated to this time in Crimea, and even farther north in the Ukrainian mainland, correspond to those of subtropical climates (Parishckura 1978; Sirenko and Turlo 1986). The Piacenzian is followed by the Gelasian, a stage that was formerly part of the Pliocene but is now in the Pleistocene (Figure 3.7).

Globally, the Gelasian stage is characterized by staggered cooling represented by several glacial advances. The more subtropical flora of the Miocene and Pliocene (Parishkura 1978; Sirenko and Turlo 1986) began to give in to mid-latitude, steppic vegetation. This part of the early Pleistocene in the regional chronostratigraphy is known as the Akchagyl stage, but it has also been divided into the Kuyalnik and Taman stages (Matoshko, Gozhik and Semenenko 2009). During the early stages of the Pleistocene, the Crimean Plains, already a terrestrial environment, began to undergo alluvial deposition and accumulation of loess (Vyeklich and Siryenko 1976). Marine regressions and transgressions occurred in tandem with the glacial and interglacial periods, which in turn created interconnections with other basins (Azov Sea–Manych–Caspian Sea) (Matoshko, Gozhik and Semenenko 2009).

The early part of the Pleistocene is known in Crimea as the Chaudian stage, named after Cape Chauda, on the south coast of the Kerch Peninsula, where marine terraces formed during marine transgressions. In more recent chronostratigraphic schemes, the Chaudian encompasses the Gurian and Chaudian proper, both of which span from about 1.8 Ma to about 0.7 Ma (Matoshko, Gozhik and Semenenko 2009). The past 0.7 million years correspond to the Chaudian stage (or Middle Pleistocene) and Upper Pleistocene stage, which are periods discussed in Chapter 5.

Modern geomorphic dynamics

Landform changes at present seem static in comparison to the evolution of the landscape through the millions and thousands of years of geological time, during which time internal (tectonic) and external (erosional) processes shaped and re-shaped the landscape. The internal processes seem slow on decadal and centennial scales but sometimes can be sudden and destructive. The Yalta earthquake of 1927 is a reminder of the active tectonic forces in Crimea. The external processes are more visible and faster in human terms and from a human perspective, and in Crimea they have been considerable on centennial and decadal scales, particularly along the coasts and in the mountainous areas, where slums, landslides, and mudslides are a common phenomenon (Klyukin 2005; Mikhaylov 2009b). Although with less frequency, these mass wasting events also occur in the valley slopes of the northern side of the Crimean Mountains area, where there is evidence of massive landslides in recent millennia (Pánek et al. 2008).

Surface erosion is a process that has been intense in all of the mountain and piedmont areas, as well as hilly areas in the Tarkhankut and Kerch peninsulas. In particular, areas of Triassic-Jurassic flysch rocks along the south coast rill and gully erosional processes have been persistent (Kyiukin 2005). Similarly, rill and gully erosion is apparently intense on the steeper slopes of the cuestas in the Inner and Outer ridges of the piedmont.

Coastal dynamics are directly related to wave action, particularly when such action is intensified by storms (Zyenkovich 1958). Sea-level change, whether occurring as a result of tectonic movements or global sea-level fluctuations, moves the level of wave erosion up or down. The formation of arches and isolated stacks in headlands on rocky coasts is evidence of processes involving waves and currents, as can be seen in many spots of the coast, such as Mys Fiolyent, on the southern part of the Heraklean Peninsula, and Maliy Atlyosh, on the southwest coast of the Tarkhankut Peninsula.

Fluvial dynamics constitute processes of erosion and sediment accumulation in river channels and floodplains. Frequent floods, sometimes of great destructive power, have occurred in historical times in the river valleys of the mountains and piedmont (Rukhlov 1915). But destructive floods along the Crimean stream valleys have been minimized with the construction of dams.

Hydrology

The largest river basins in the Crimean Peninsula are on the northern slopes of the mountains and the piedmont (Figure 1.2). Smaller river basins with large gradient streams (torrents) characterize the southern slopes of the mountains. The plains, on the other hand, are crisscrossed by large rivers or drained by smaller low-gradient streams. The yailas, being mainly a karstic terrain, lack streams; most of the rainwater either seeps into the porous rock or collects in small basins (Olifyerov 2011).

Crimean river basins are often subdivided into major drainage areas: the western watershed, or rivers flowing into the Gulf of Kalamita; the eastern watershed, whose rivers flow into the Sivash; the southern watershed, or rivers draining the south slope of the mountains; and the plains watershed. Data provided below were published by Pyotr D. Podgorodyetskiy (1988) and A. N. Olifyerov and Z. N. Timchenko (2009).

The major rivers of the western watershed are the Chyornaya, Bel'bek, Kacha, Al'ma, and Bulganyak (Figure 1.2), all of which empty into the west coast (i.e., the Gulf of Kalamita). Of the four, the Al'ma is the longest river (84 km), covering an area of 645 km^2 and with an average discharge of 1.21 m^3/sec. The highest average discharges in this watershed, however, are produced the Bel'bek (2.75 m^3/sec) and Chyornaya (1.94 m^3/sec), which are, in fact, the two largest in Crimea. Along their course through the lowest part of the piedmont and the plains, the western watershed streams form narrow floodplains, often bordered by several flights of alluvial terraces, which mark former floodplains abandoned by the river as it erodes down. The Chyornaya River, however, is different from the rest of the large rivers, because it is basically a mountain stream along almost its entire path. The Chyornaya flows from the western edge of the Ay-Petrinskaya Yaila into the Baydar Valley, where it is fed by the Uzundzha River before being dammed; from the Baydar Valley it descends along another canyon into another dam, and then into the Inkerman Valley, where it forms a broad floodplain before it empties into Sevastopol Bay. Its lower floodplain is very deep and contains a variety of sediments including some lacustrine and lagoonal facies (Cordova et al. 2011).

The eastern watershed streams are the Salgir, Mokriy Indol and Chorokh-Surivers. The Salgir River has the greatest length (238 km) and largest basin (4,010 km^2) of all Crimean rivers. It collects water from large tributaries, such as the Bulganyak Vostochny and the Karasu rivers, which drain large tracts of the central and western parts of the south slope of the mountains. With an average discharge of 1.8 m^3/sec, the Salgir River ranks third in average discharge after the Bel'bek and Chyornaya rivers.

The streams of the southern watershed include mainly short streams draining small and very steep basins. From west to east the main basins include the following rivers: Uchan Su, Dyeryekoyka, Ulu-Uzen', Demerdzhi, and Vostochnyy Ulu-Uzen'. The latter is the longest river in this region, with a length of 16 km. The largest basin corresponds to the Ulu-Uzen' (60.8 km^2), which drains part of the Angara Pass and empties into the sea near Alushta. The highest average discharge corresponds to the Dyeryekoyka (0.51 m^3/sec), which is only 12 km long and has a basin covering 44.3 km^2.

The plains watershed is drained by small streams that are not connected to the large rivers of the western and eastern watersheds. The streams of the Tarkhankut and Kerch peninsulas are controlled by synclines and eroded anticlines and other tectonic structures. Although small, the largest streams of the plains watershed are in the Tarkhankut Peninsula: the Chatyrlyk (132 km long) and the Samarchyk (44 km long). Although their basins are quite large, 2,370 km^2 and 484 km^2, respectively, their discharge is low and intermittent (i.e., seasonal) due to their location in an area of low precipitation. The Kerch Peninsula has even smaller streams, all of which are intermittent.

Discharge in Crimean streams varies considerably throughout the year, responding in large part to the availability of rainwater and the effects of evaporation, and to a lesser degree to the feeding of underground water. Although damming and water tapping has created variations in the hydrological regime, the natural stream discharge increases during the winter as rains and snowmelt produce water for recharging aquifers, reaching a peak in March (Olifyerov and Timchenko 2009). Throughout the summer months, as rains decrease and evaporation rates increase, discharge in the rivers diminishes. Limited feeding from underground water starts to diminish as early as June, reaching its minimum in August and September (Olifyerov and Timchenko 2009).

Crimea's natural lakes are concentrated in the northern plains near the Sivash and Perekop areas, the Kerch Peninsula and along the coasts (except along the coast south of the mountains). The only natural lacustrine systems in the mountains are small ephemeral lakes or ponds in the Karabi and Demerdzhi yailas. These ponds are formed in karstic depressions that have been sealed with clay produced by weathering of surrounding rocks (Olifyerov and Timchenko 2009).

Practically all the lakes in the plains are shallow and their waters are saline. Their maximum depths range between 0.3 m and 2 m, with the exception of

Lake Donuzlav, whose maximum depth is 27 m (Yena, Yena and Yena 2004b). The coastal lakes are referred to as liman lakes, which act more as lagoons since some of them exchange water with the sea. The term *liman*, which in its original Greek form meant harbor, is used in the local geomorphological lexicon to refer to a valley flooded by rising seas during the Holocene (Podgorodyetskiy 1988).

The high salinity of the Crimean lakes of the plains is due to high evaporation rates and, in the case of the coastal lakes, to the incursion of sea water (Dzyens-Litovskaya 1970; Blagovolin 1975). In fact, salinities in the coastal lakes usually reach 250–270 permil (Yena, Yena and Yena 2004b). Salt procurement from the saline lakes has been the traditional activity of the Crimean Plains, and has remained so to this day. The use of curative muds from the lakes is a further addition to therapeutic tourism, as evidenced by the sanatoria around Lake Saki.

Coasts, lagoons and natural harbors

Coastal landforms are influenced by lithology, tectonics, the depth of the sea, the influence of incoming river sediments, and exposure to winds and storms (Zyenkovich 1958). Erosional coasts are characterized by cliffs and narrow pebbly beaches, as is the case of most of the southern coast and parts of the Tarkhankut and Kerch peninsulas. In erosional coasts, tectonic activity, constant landslides, which cause steep slopes, and exposure to storms are the main geomorphic modifiers of the coast. Accumulative coasts correspond to sandy beaches, sand bars, and in the case of some parts of Crimea, shell-fragment beaches and bars. Sandy beaches characterize the shores of the Kalamita Bayand some bays in the Kerch Peninsula. Sandy and shell-fragment beaches also characterize the shores of the Arabatskaya Strelka, a sandy bar that separates the Sivash lagoon from the Azov Sea (Figure 1.1). The coast of the Azov Sea presents other characteristics, principally in the formation of cumulative coastal landforms like sand spits and a *tombolo* (e.g., the land neck of Cape Kazantip) and sand bars (e.g., the Arabatskaya Strelka) (Klyukin and Korzhyenyevskiy 2008).

Lagoonal systems in Crimea are often associated with the liman lakes mentioned above and with very shallow bodies of water separated from the sea by sand bars. The best examples of lagoons are Lakes Saki and Lake Yevpatoria along the southern coast of the Tarkhankut Peninsula, and Lake Uzunlar and Lake Tobichyesk on the eastern shore of the Kerch Peninsula. The largest lagoonal system in Crimea is the Sivash, known in Russian also as *Gniloye Ozyero* (the putrid sea, due to its unpleasant odor), a hypersaline body of water that today forms most of the northern flank of the Crimean Peninsula (Figure 1.2)

The Sivash is a system of interconnected lagoons occupying an area of 2,540 km^2. Stashchuk et al. (1964) divided the Sivash into western, central, eastern, and southern basins (Figure 1.2). The depth of the Sivash varies from less than 0.5 m in the western basin to about 3 m in the southern basin. The central and eastern basins have an average maximum of 1 m and 2 m, respectively. The Sivash's salinity is primarily the result of high evaporation rates, which are around 1,100 mm a year (Stashchuk et al. 1964)—in fact more than three times the amount of rain a year. Salinity in the Sivash can vary throughout the year, reaching concentrations as high as 270 permil (Stashchuk et al. 1964). But salinity varies also across the system, with the lowest values in the eastern basin, where the system receives the direct flow of water from the Azov Sea, whose salinity is only 11 permil, via the strait of Genichyesk. Salinity increases from this point, reaching its highest values in the strait connecting the western and central basins and in the southern end of the southern basin (Stashchuk et al. 1964). Rivers should reduce the salinity of the Sivash, but the Salgir River, the major stream emptying into the southern basin, no longer reaches the Sivash due to water use along its path.

Bays and natural harbors have always been important in Crimean history. The two most coveted harbors have been Sevastopol Bay and Balaklava Bay, both of which are deep, protected from storms, and ice-free in winter. Smaller and less protected bays in other parts of the Crimean coast include Sudak, Yevpatoria, Feodosia, and Kerch. Others, like Yalta, have artificial harbors. The only ports that handle large tonnage and larger cruise ships are Sevastopol, Yalta, Feodosia and Kerch. Of them, only the latter has problems with water freezing over in winter.

Surrounding seas

Despite being connected, the Black Sea and the Azov Sea contrast greatly in terms of depth, salinity, weather phenomena, water temperatures and fauna. In terms of average values, the salinity of the Black Sea is 15 permil on the northwest coast, 18 permil in the center of the basin, and 22.6 permil at depths below 1,000 m (Tamaychuk, Andreyeva and Penno 2009). Compared with the average ocean salinity (35 permil), these values are relatively low.

Water temperatures in the Black Sea vary with season and location. July temperatures vary from 28°C in the open sea to 23°C in bays. By contrast, January temperatures vary from 8°C in the open sea to 0.5°C in bays. Sea currents are primarily influenced by the action of atmospheric circulation (Kosarev, Arkhipkin and Surkova 2008). The main currents form two main cells called the western and eastern gyres (Figure 1.5).

The strong winds of the storms produce surges and waves of up to 3 m, with exceptional ones of up to 7 m in height in the open part of the sea (Kosarev, Arkhipkin and Surkova 2008). Storms, very common all over the Black Sea, have caused several maritime catastrophes, such as the sinking of 34 battleships of the Anglo-French squadron on 4 November 1854, during the Crimean War (Kosarev, Arkhipkin and Surkova 2008). In January 2001, a passenger vessel traveling between Istanbul and Sevastopol sank after being hit by a storm half way through its trip. Luckily, all passengers and crew were rescued. Storm waves can be so destructive that, in some instances, they can damage infrastructure even inside harbors. As an example, on 11 November 2005, a destructive storm hit western Crimea, producing a surge that broke into Sevastopol Harbor and damaged the west-side quays of Streletskaya Bay.

The Azov Sea, known in ancient sources as Lake Maeotis, occupies an area of 290 km^2, and has an average depth of -7 m with a maximum depth of -13 m. With the influx of fresh water from the large rivers flowing into it (e.g., the Don and Kuban), the waters of the Azov Sea are basically brackish, with a salinity of 10–11 permil. This low salinity is caused by the influx of water from the Black Sea and the Sivash. Average temperatures of the water are much lower than those of the Black Sea, ranging from 28°C to 30°C in summer and from −1°C to −5°C in winter. Its limited circulation, low salinity and northern latitudinal position lead to the formation of surface ice, presenting an obstruction to navigation during the winter months.

Weather and climate

The main factors controlling weather are solar radiation, atmospheric circulation, topography and the surrounding seas. Solar radiation is directly related to Crimea's location in the mid-latitudes (N 44°–46°), where strong differences in radiation exist between winter and summer, resulting in a broad range of seasonal temperatures and evapotranspiration rates. Modifications to the incidence of solar radiation exist due to cloudiness and topography. Cloudiness varies regularly and seasonally but, overall, clear skies predominate throughout the year, allowing land and water to warm up faster than do regions in western Europe (Kosarev, Arkhipkin and Surkova 2008). Topography influences the incidence of solar radiation, depending on the orientation of the slope. Thus, north-facing slopes and slopes in narrow valleys tend to receive less solar radiation.

Atmospheric circulation over Crimea is highly dependent on seasonal developments of the Polar Jet Stream and the Eurasian High. The Polar Jet Stream—a high-altitude westward wind flow—works in tandem with the lower-level westerly winds in driving rotating transient systems known as mid-latitude cyclones. Mid-latitude cyclones, or low-pressure systems that carry cold fronts, produce rainfall (Kostopolou 2007). These systems move across Europe or across the Mediterranean, sometimes making it north towards Crimea, where normally cold air forms fronts that also produce rain (Kostopolou 2007). As a consequence of this cyclonic path, about 75 percent of all precipitation in Crimea originates in the Atlantic Ocean, 15 percent from the Mediterranean Sea, and 10 percent via cold fronts (Vyed' 2000). How much originates in the Black Sea is not known, but it is possible that the cyclones moving from the west (Atlantic) or from the southwest (Mediterranean) pick up moisture from the relatively warm waters of the Black Sea (Yergina 2009). Storms intensify between October and January, during which time the westerly flow and the southerly excursions of the Polar Jet Stream are more frequent.

The Eurasian High is an anticyclone (descending air circulating clockwise) that acts as a major gear for the movement of winds clockwise and in blocking the path of the westerly winds and mid-latitude cyclones (Vyed' 2000; Yergina 2009). Thus, when the Eurasian High expands, it blocks the westerlies, producing dryness over Crimea. A strong gradient from low pressure in the west and high pressure in the east normally sets in (Vyed' 2000). During the winter time, incursions of Arctic air behind cold fronts bring below-freezing temperatures.

Local topography modifies the effects of atmospheric circulation by increasing or reducing the amount of precipitation on the land. Accordingly, the mountains receive more precipitation because air is forced to rise and produces more precipitation. The plains, on the other hand, receive less precipitation (Figure 3.10A). The Crimean Mountains produce a *foehn* wind effect that creates descending warm dry winds from the mountains onto the south coast (Kosarev, Arkhipkin and Surkova 2008). This effect results in the constant tablecloth-like clouds over the mountains during the summer, which disappears at the southern edge of the yailas (see mountain background in Figure 1.5). This effect produces dryness on the south coast during summer.

Essentially, the mountains protect the south coast from the blasts of Arctic air. This effect helps maintain relatively mild temperatures south of the mountains (Figure 3.10B), which is one of the reasons for the development of subtropical vegetation in this region. Thus, while the mean monthly temperatures in Simferopol plunge below freezing, Yalta maintains temperatures slightly over 0°C (Figure 3.10C). The winter temperature minima in Simferopol can drop to −12°C or lower, while in Yalta they rarely fall below −3°C (Figure 3.11). Another effect of the local topography is on the sea breezes, which tend to be more intense where the highest contrast between land elevation and the sea surface exists. This contrast is nowhere

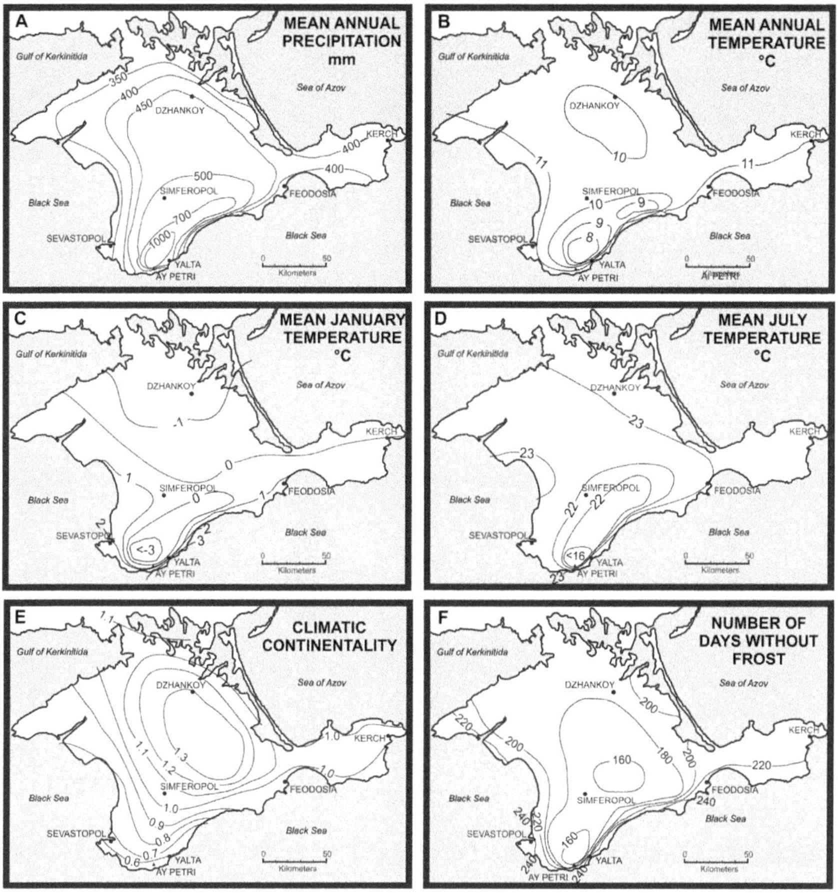

Figure 3.10: Basic climatic maps. Source: Vyed' (2000).

else in Crimea as intense as along the south coast (Kosarev, Arkhipkin and Surkova 2008).

Because land surfaces respond faster to atmospheric temperature changes than water surfaces do, temperatures in mainland interiors can reach extreme highs and lows. This effect is known as continentality because it is more common in areas away from the sea. The water surrounding Crimea reduces the effects of continentality, making winters milder and summers cooler than in the southern steppes of mainland Ukraine. However, minimal effects of continentality are experienced in the east-central part of the plains (Figure 3.10E).

Because most mid-latitude cyclones move from west to east, or southwest to northeast, the areas benefiting from more moisture are in the southwestern part of the peninsula. This effect is seen in the gradient of mean precipitation, which is higher in the western parts of the mountains and piedmont (Figure 3.10A). Thus, the summits of the mountains in the west receive more annual

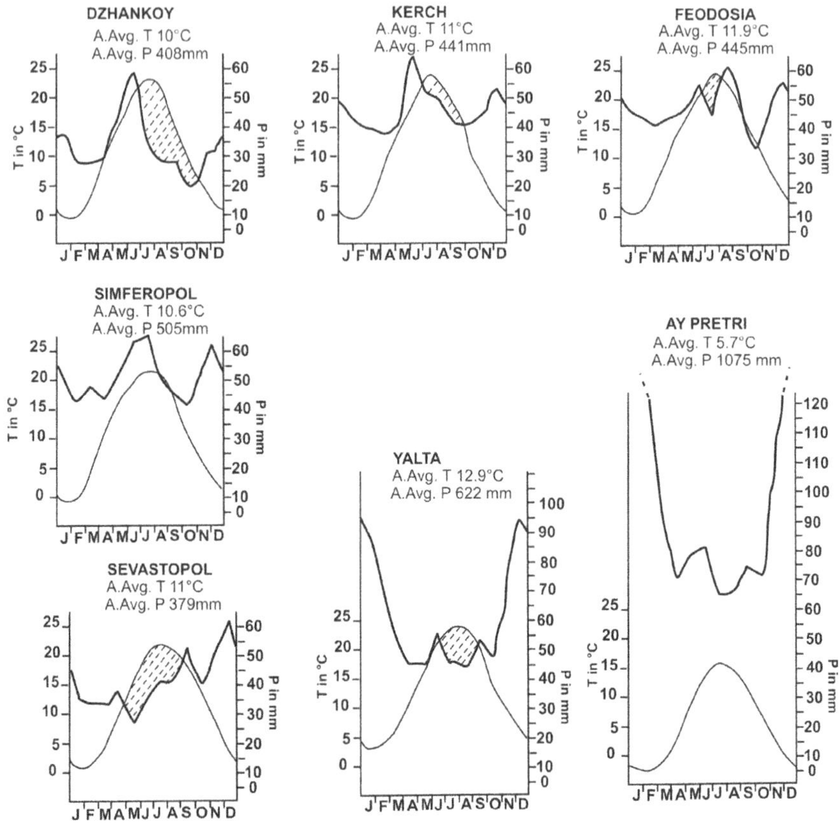

Figure 3.11: Climographs. Source: Cordova et al. (2011).

precipitation than those in the east. Compare for example the 1,075 mm recorded in Ay Petri (in the western yailas) versus values around 700–800 mm in the eastern yailas. Likewise, a precipitation gradient is observed on the south coast between Yalta (622 mm) and Theodosia (445 mm).

The Azov Sea contributes to rainfall over Crimea, though minimally. During July and August, when the east–west gradient of atmospheric pressure is reduced and conditions become calmer, temperatures in the plains rise, creating a contrast with the Azov Sea, and consequently attracting breezes that can carry over and feed moisture for thunderstorms (Vyed' 2000). Also during the summer, orographic rain occurs on the northern slopes of the mountains (Yergina 2009). This phenomenon obviously does not occur on the south coast, as the mountains separate it from the conditions taking place in the rest of the peninsula.

The climate of the south coast, with its relatively mild winters and dry summers, is often referred as Mediterranean, or more properly as Sub-Mediterranean. The relatively dry summer and the relatively small number of days with frost compared to other parts of Crimea (Figure 3.10F) make this region's climate similar to that of the Mediterranean. However, Douguedroit and Zimina (1987), who compared the climatic characteristics of the south coast of Crimea with the Mediterranean, found two main differences: the Crimean winters are much colder than those in most of the Mediterranean region, and the summers are not as quite as dry as in the Mediterranean. A short period of dryness in the summer is often interrupted by a spell of rain, often in June (see Ay Petri and Yalta diagrams in Figure 3.11). Yet, the climate is warm enough to sustain Mediterranean flora, native and introduced. Interestingly, the closest matches between Crimea's south coast and the Mediterranean are the mountainous areas of southern France or northern Italy (Douguedroit and Zimina 1987).

The northern slopes of the mountains experience lower winter temperatures and overall a higher mean annual precipitation (above 700 mm). The plains region receives lower amounts of precipitation (often less than 450 mm annually) and has higher rainfall variability than do other parts of Crimea (Kochkin 1967). Winter temperatures on the plains and piedmont are extremely low compared to the southern slopes of the mountains (Figure 3.10C). Average January temperatures in the plains and piedmont are of the order of from 0°C to −2°C, whilst absolute minimum temperatures can be as low as −30°C in Dzhankoy and Simferopol (Vyed' 2000). By contrast, winter temperatures on the south coast are of the order of 3–4°C, with an absolute minimum temperature of −15°C registered in Yalta. Summer temperatures on the plains are similar to, or slightly higher than, those on the southwest coast (Figure 3.10D). Thus the south coast, in addition to its relatively warm temperatures in winter and summer, also experiences less extreme temperatures through the seasons.

Soils

Due to the diversity of landforms, geological substrates and hydrological features, soil types vary considerably across the Crimean Peninsula (Figure 3.12). The characteristic soils of the plains are the salinized types in the north, and the steppe black earth soils in the south. In the piedmont and mountains, the main types include stony soils associated with the calcareous rocks and forest soils. The south coast, although part of the mountains, exhibits remarkably different soil types, mainly the Sub-Mediterranean red types.

In the main, the soil terminology used in Ukraine, as well as in Russia and other former Soviet Republics, has its origins in Vasiliy V. Dokuchayev's classification (created in the late 1800s), which has been modified over time. Unlike the 7th Approximation (the classification used in the United States)

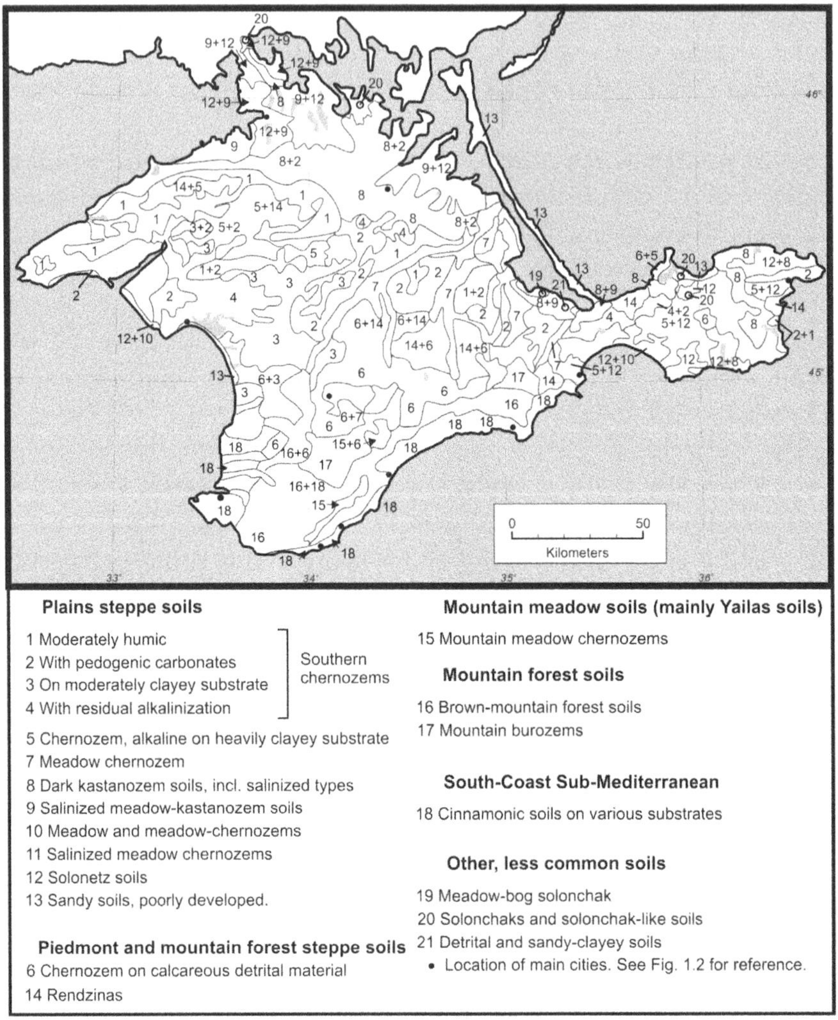

Figure 3.12: General soil types. Source: Dragan (2003).

and the widely used FAO-UNESCO, the classification used in the former Soviet Union has a very strong emphasis on soil genesis and climate. In recent years, however, Natalia N. Dragan (2000, 2003, 2009) has revised and adapted the classification to the local conditions of Crimea (Figure 3.12).

The northernmost parts of the plains, mainly around the Sivash, are dominated by the highly alkaline and salinized soils, or Solonetz types (Figure 3.12). Natural soil salinization is due in part to a high water table (hydromophic factor) coupled with high and low mean annual precipitation and evapotranspiration rates (climatic factors) (Dragan 2009). A gypsic horizon is often evident in these soils (Figure 3.13). Towards the south, these

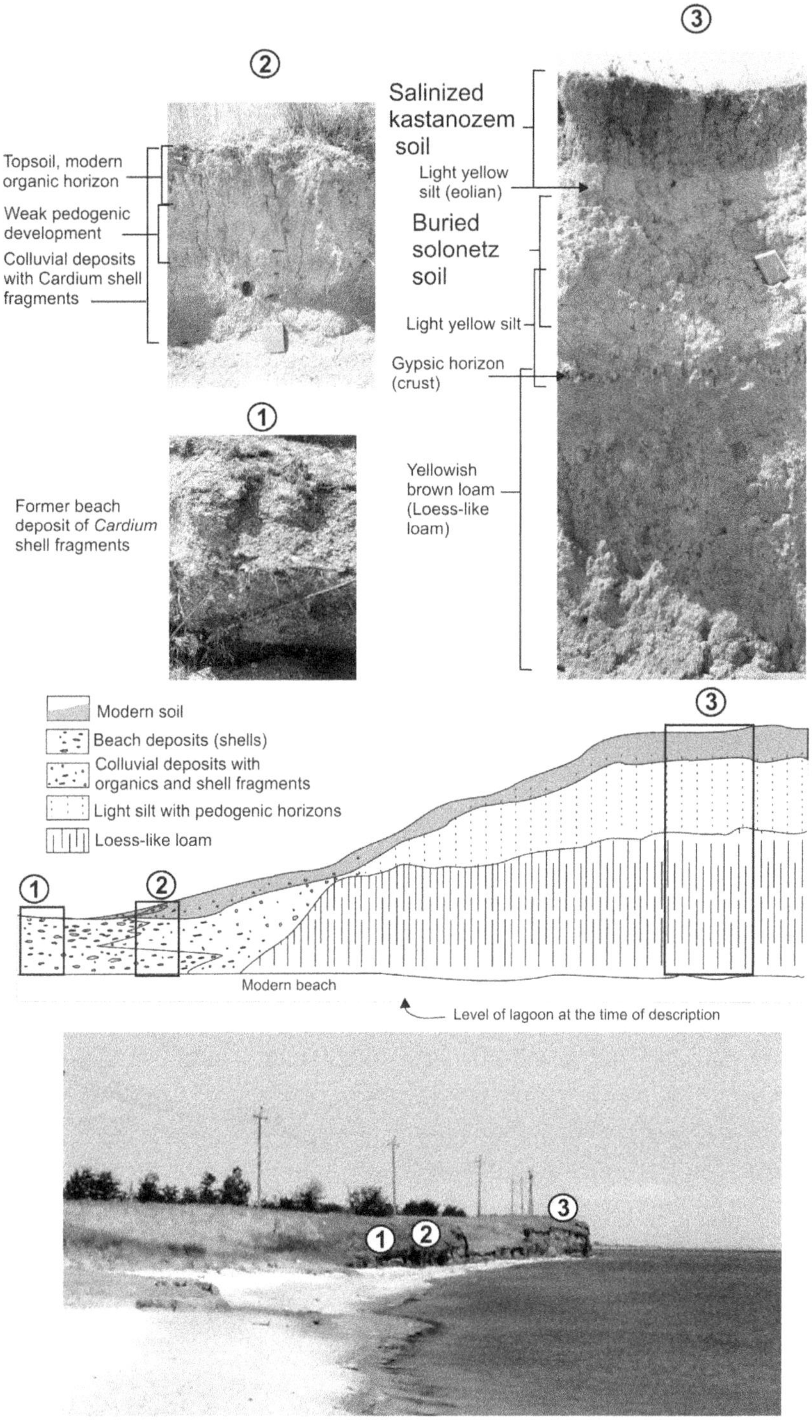

Figure 3.13: Sivash: Stratigraphic section and soil profiles.

soils grade into salinized kastanozems (also known as chestnut soils) as salt concentrations decline and soils present more organic matter (Dragan 2009). Very commonly, the kastanozems are developed on loess and loess-like deposits of Late Pleistocene age (Blagovolin 1975). This is the same loess deposit as the one shown in the profile of Figure 3.13. The kastanozems are often salinized, particularly in the north, but farther south they become richer in humus and consequently are referred to as dark kastanozems.

The dark kastanozem soils grade into chernozems (black earths), which are soils characterized by strong humification represented by a dark horizon, very typical of grassland vegetation that undergoes long cold winters and precipitation amounts lower than in forested areas. However, it is important to point out that the Crimean chernozems differ from those of the Ukrainian and Russian steppes in that they present accumulation of pedogenic carbonates in a subsoil layer. These soils are defined as "southern chernozems" or, more properly, "Crimean southern chernozems" (Krupskiy and Polupan 1979).

The pedogenic carbonate horizon in the southern chernozems, or Bk horizon, is common also in other soil types in the Crimean Peninsula. In the center of the plains and in some areas on the west coast and the Kerch Peninsula, chernozems are formed on loess and loess-like clayey deposits. As the plain rises on the piedmont and soils become more influenced by the forest-steppe vegetation, or less dry conditions, they become less salinized and more humified (Krupskiy and Polupan 1979; Dragan 2000, 2009).

As the piedmont turns into steeper mountainous slopes, chernozem soils grade into rendzinas (thin dark humic soils in a stony calcareous matrix) and meadow chernozems. As forests become denser higher in the mountains, soils become of the mountain brown forest type and brunizems. These are soils with high amounts of organics and no pedogenic carbonates (i.e., no Bk horizon).

Soils in the Yailas are characterized by rendzinas and meadow chernozems. The rendzinas are developed under petrophytic steppe conditions. The meadow chernozems are usually formed in depressions, where sediments are deeper and soil moisture is greater, hence forming wet meadows, or the *feucht wiese* of Walter (1974), which are communities dominated by grasses, sedges and a diversity of forbs. The southern slopes of the mountains are characterized by bare rock and, in some cases, where topography allows it, by rendzinas.

The southern coast between Kara-Dag and the Heraklean Peninsula is dominated by the cinnamonic soils (*korichenevie pochvy* in Russian). The cinnamonic soils were identified by S. A. Zacharov (1929) and popularized in the USSR soil literature by Inokyenty P. Gerasimov (1954), who defined them as subtropical soils, typical of the dry forests, which implies also some affinities with the red Mediterranean, or *terra rossa*, soils. In the *Atlas of Soils* of the Ukrainian SSR, they are defined as "red cinnamonic soils of dry forests and shrublands" (Krupskiy and Polupan 1979). It is not clear whether the processes of formation of the cinnamonic soils of Crimea are the same as those that formed the typical red Mediterranean soils or the *terra rossa* proper (Dragan 2009), but in Crimea the cinnamonic soils occupy the areas with maximum precipitation in winter and the lowest incidence of frost, and with a strong presence of Mediterranean floristic elements.

Depending on the substrate, the amount of humus in their profiles and their development, three variants of cinnamonic soils are distinguishable: typical, non-calcareous, and calcareous (Dragan 2009). The typical cinnamonic soils are characterized by well-differentiated horizons:

A horizon: 25–35 cm thick humic horizon
Bt horizon: a conspicuous illuvial horizon
Bk horizon: a pedogenic carbonate horizon

They are conspicuously developed on Mio-Pliocene sedimentary units of southwestern Crimea (Dragan 2009; Lisetskii et al. 2012) and sediments in depressions and *balki* (ravines) in the Heraklean Peninsula (Cordova and Lehman 2005). The non-calcareous cinnamonic soils are developed on Triassic-Jurassic shale on the south coast, particularly between Alushta and Feodosia. Their humic horizons are considerably thicker (up to 70 cm) and their lower horizons poorly differentiated. The calcareous cinnamonic soils are developed mainly on marl and limestones and calcareous conglomerates of various ages. These soils are poor in clay, rich in non-pedogenic calcium carbonates, and their horizons are poorly differentiated. In general, these soils are a transitional type between cinnamonic and rendzina soils.

CHAPTER 4

FLORA AND FAUNA

The flora of Crimea

The best way to look at the peninsula's flora is through an overview of the most common species, grouped into life forms—trees, shrubs, forbs and graminoids—and followed by a brief description of the main plant communities (associations) that they form. For that purpose, this description refers to the main vegetation regions of the peninsula (Figure 4.1), their altitudinal distribution in the mountains and piedmont (Figure 4.2), and a proxy reconstruction of plant communities that have been lost under the plow in the plains (Figure 4.3).

Trees and shrubs

Woody species in Crimea are distributed according to altitude belts (Table 4.1; Figure 4.2). This distribution is the result of different tolerances to extreme temperatures, drought and wind (Vyed' 1983), as well as factors such as slope orientation and protection in valleys (Shelyag-Sosonko, Osychnyuk and Andriyenko 1982). The most common trees and shrubs correspond to those that have dominant or edificator roles in their vegetation communities, as is the case, for example, of beech in the beech forest, or pistachio and juniper in the pistachio–juniper woodland. Many other species may not be dominant in the communities but they are widespread through the landscape, which is a reason why they deserve to be mentioned.

Pine trees are concentrated mainly on the southern slope of the mountains (Figure 4.2), on some slopes of the western part of the mountains, and on the yailas. They are also abundant in other areas of the peninsula where they have been deliberately planted. The three native species of pine in Crimea are *Pinus pallasiana*, *Pinus sylvestris* subsp. *kochiana*, and *Pinus pityusa*.

Pinus pallasiana (Crimean pine) is believed to be a subspecies of *Pinus nigra* (black pine) with close relatives in the Caucasus and northern Turkey

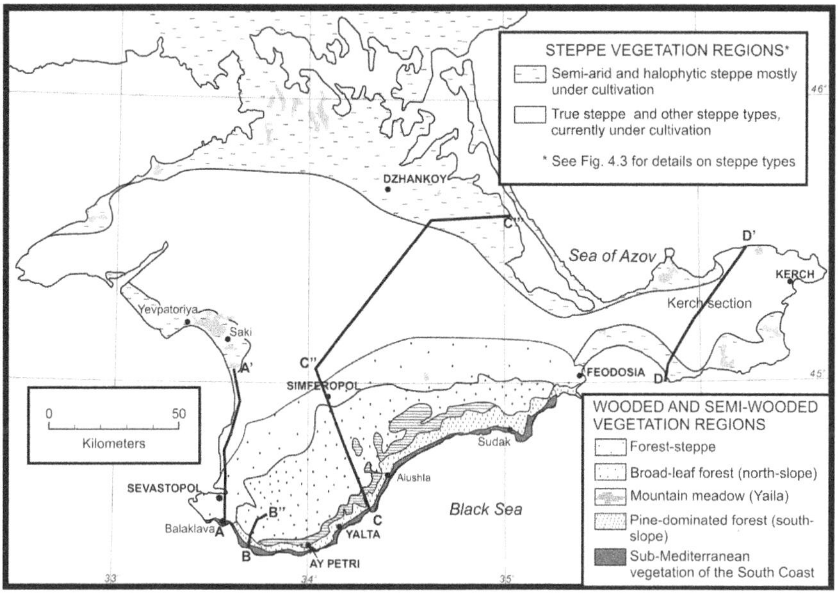

Figure 4.1: Vegetation units. Source: Biodiversity Support Program (1999) and Cordova et al. (2011), with modifications.

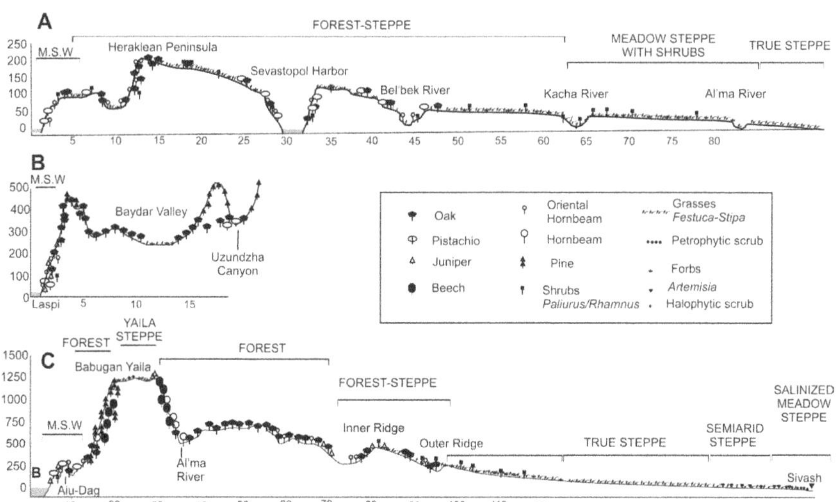

Figure 4.2: Vegetation transects across (A) the Heraklean Peninsula and west coast; (B) the south coast – Baydar Valley – Uzundzha Canyon; (C) the mountains, piedmont and plains.

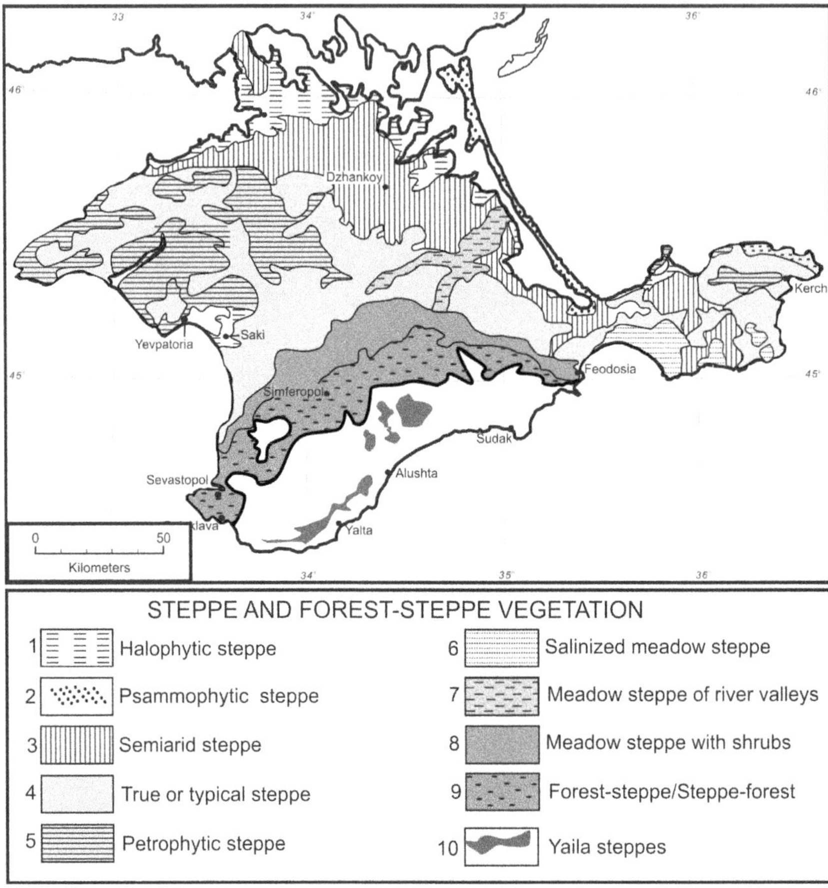

Figure 4.3: Original vegetation communities in the steppe and forest-steppe. Compiled from various sources: Rubtsov (1978), Podgorodyetskiy (1988), Didukh (2003).

(Eckenwalder 2009). In Crimea *P. pallasiana* is widespread in the forests on the southern slopes of the mountains particularly between 600 m and 1,100 m of elevation. It is also found on the slopes of the Baydar Valley and adjacent canyons (Figure 4.2B).

Pinus sylvestris subsp. *kochiana*, a variety of Scot's pine (*P. sylvestris*), is found in the Transcaucasian region (i.e., eastern Georgia and Armenia), northeast Turkey, and Crimea (Eckenwalder, 2009). In the botanical literature, this pine is often referred to as *Pinus kochia* (i.e., Shelyag-Sosonko, Osychnyuk and Andrienko 1982; Vakhrushyeva 2009). In the Crimean Mountains, Koch's pine is found in the highest forest belt, predominantly on the southern slopes and often around the yailas (Table 4.1; Figure 4.2), where it grows short and with bent trunks due to the constant strong winds.

Table 4.1: Altitudinal ranges of arboreal species of the Crimean Mountains and Crimean Piedmont (after classification by Didukh 1992 and modifications by Cordova, Rybak and Lehman 2001).

Formation	**Elevation range (in meters above sea level)**	**Location**
Greek/tall juniper (*Juniperus exelsa*)	0–450, rarely to 600	1 (mostly on western end), 4
Pubescent oak (*Quercus pubescens*)	0–450, rarely to 600	1, 4
Oriental hornbeam (*Carpinus orientalis*)	0–450, rarely to 600	1, 4
Bluntleaf pistachio (*Pistacia mutica*)	0–250	1 (only on western end), 4
Stankewicz's pine (*Pinus brutia* var. *pityusa*)	0–250	4
Durmast oak (*Quercus petrae*)	450–500, rarely to 800	2, 3
Hornbeam (*Carpinus betulus*)	500–1,100	2, 3
Tall ash (*Fraxinus excelsior*)	500–900	2, 3
Crimean pine (*Pinus pallasiana*)	500–1,200	3
Koch's pine (*P. sylvestris* var. *kochiana*)	900–1,300	3, 5 (rare)
Beech forests (*Fagus sylvatica*)	600–1,300 (northern slope) 800–1,300 (southern slope)	2, 3 (less common)
Steven's maple (*Acer steveni*)	1000–1,300	2
Sticky alder (*Alnus glutinosa*)	500–1,000	2 (restricted to some valleys)

(1) piedmont; (2) mountains, northern slope; (3) mountains, southern slope; (4) south coast belt; (5) yailas

Pinus brutia var. *pityusa* (Farjon 2001), or Stankiewicz's pine, and its Caucasian close relative *Pinus brutia* var. *pityusa* (Farjon 2001), are considered two disjunctive northern varieties of *P. brutia* (Turkish pine) (Conkle, Schiller and Grünwald 1988; Goncharenko et al. 1998; Farjon 2001; Yena 2007, 2012). This genetic relation has resulted in the removal of Stankiewicz's pine from the list of endemic species of Crimea (Yena 2007).

In Crimea, Stankiewicz's pine is restricted to three localities on the south coast: one in the west near Cape Ayya, another in the east in the Sudak-Novy Svyet area (depicted in Figure 4.6), and one discovered in 1995 on the slopes of Mount Papaya-Kaya above Cape Ay-Foka, 4 km due west from Novy Svyet (Yena, Yena and Yena 2005). In all three cases this pine species grows on detrital soils on steep slopes (50–60°) and at elevations no higher than 200 m. Additionally, a plantation of this pine, some 55–60 years old, exists in the area of Morskoye, about 10 km west of Novy Svyet (Goncharenko et al. 1998).

Oak is the most abundant tree in the forested parts of Crimea, where there are three native species: *Quercus pubescens*, *Q. petrae*, and *Q. robur*. The pubescent oak (*Q. pubescens*) is a southern European species, although it can be found in some forests in central and eastern Europe and Anatolia (Davis 1982; Flora Europaea Database 2012). In Crimea, *Q. pubescens* is the dominant species in the lower forest belts of the northern and southern slopes of the mountains, the *dubyanik* vegetation of the piedmont, and the *shiblyak* vegetation of the south coast, where it forms associations with *Carpinus orientalis*, *Pistacia mutica* and *Juniperus excelsa* (Didukh 1992; Vakhrushyeva 2009).

The sessile oak (*Q. petraea*) and the cherry oak (*Q. robur*) have a broader distribution in Europe, Anatolia and the Caucasus but are less frequently found in the Mediterranean region (Flora Europaea Database 2012). In Crimea, both species are found in the mountains at elevations usually above 400 m (Rubtsov 1972, 1978). It has been proposed that *Q. robur* may have been dominant in certain communities but, because of human-led destruction, it has been replaced by other trees (Vakhrushyeva 2009).

Hornbeam in Crimea is represented by *Carpinus betulus* and *C. orientalis*, each occupying different altitudinal belts (Table 4.1). *Carpinus betulus* (European hornbeam) is common in the temperate forests of central Europe, the Caucasus, and other mountainous regions of western Asia (Didukh 1992). In Crimea it is found mainly at elevations above 600 m, particularly on the northern slopes, where it forms associations with beech (*Fagus sylvatica*) (Rubtsov 1978; Didukh 1992). *Carpinus orientalis* (Oriental hornbeam) extends from the Balkans, through Crimea, the Caucasus, Anatolia, and northern Iran (Browicz and Zieliński 1982–94, vol. 3). It is a small tree (10 m tall maximum) that in Crimea often grows to a tall shrub size and in co-dominance with *Q. pubescens* and *Pistacia mutica* (Didukh 1992).

Beech is a dominant tree in the high altitudinal levels of the Crimean Mountains. Although the names *Fagus sylvatica*, *Fagus orientalis*, *Fagus sylvatica* var. *orientalis*, and *Fagus taurica* have all been proposed, there is still a debate regarding the taxonomic definition of the beech species in Crimea (see below). Therefore, for the time being, the name *Fagus sylvatica* is used here. The Crimean beech occupies an altitudinal level between 600 m and 1,000 m on the northern and southern slopes of the mountains, where it can be dominant or co-dominant with *Carpinus betulus* and *Pinus sylvestris* var. *kochiana* and *P. pallasiana* (Shelyag-Sosonko, Osychnyuk and Andriyenko 1982; Didukh 1992). Beech trees are also found on the yailas, where they only attain a short height (less than 2 m) because of to the strong wind and cold (Wulff 1932).

As mentioned, the taxonomic definition of the beech in Crimea has been somewhat problematic and it remains inconclusive. The high leaf polymorphism found in Crimean beech specimens suggested that the Crimean beech trees are the result of the hybridization of *F. sylvatica* and *F. orientalis* (Wulff 1932; Didukh 1985). Moreover, the apparent isolation of the Crimean beech community from the rest of the Eurasian communities, and the role of the Crimean Mountains as a glacial *refugium*, led to the idea that the Crimean beech should be considered a separate species: *Fagus taurica* (Didukh 1985). A more recent study has confirmed the genetic separation of Crimean beech trees from the rest of the Eurasian *Fagus* species, supporting the indication of hybridization and strengthening the idea of geographic isolation (Gömöry, Paule and Vyšný 2007). Although the role of the glacial refugia seems to have been important in this process, during the Holocene the Crimean beech remained isolated from other communities in Europe, which increased their chances for allopatric speciation (Gömöry, Paule and Vyšný 2007).

Pistachio trees in Crimea correspond to *Pistacia mutica* (blunt-leaf pistachio), which is often assigned to a subspecies of *Pistacia atlantica* (Davis 1982), a species widespread in the eastern Mediterranean and some of the mountainous areas of southwest and central Asia (Browicz and Zieliński 1982–94, vol. 3). In Crimea, *P. mutica* has the northernmost location of the species and, indeed, of the entire genus *Pistacia* (Vodop'yanova 1986), which is why it is an important Mediterranean element in Crimea and an indicator of mild winter temperatures (Cordova and Lehman 2005; Cordova 2007). Peter S. Pallas (1812) believed that the bluntleaf pistachio was introduced to Crimea by the Greek colonists in ancient times, a belief that remained rooted among scholars for a long time (Zgurovskaya 1996). However, some botanists considered that the Crimean pistachio was already part of the native Mediterranean (or Sub-Mediterranean) flora of Crimea (i.e., Grosset 1979a, 1979b). In the pollen diagrams of the Heraklean Peninsula, *Pistacia* pollen appears as early as 7,500 years BP (Cordova and Lehman 2005). Evidence of pistachio trees in previous periods, particularly during the Pleistocene, has not been found. The survival of pistachio trees in glacial refugia in lower areas on the south coast—at much lower levels than today—remains a hypothesis to be tested. Postglacial re-colonization from the Black Sea's southwestern shore via marine currents is another hypothesis (Cordova 2007).

Pistacia mutica in Crimea forms associations with *Quercus pubescens*, *Carpinus betulus* and *Juniperus excelsa* in the shiblyak vegetation community of the south coast and the Heraklean Peninsula (Didukh 1992). Although widespread along the south coast between Feodosia and the Heraklean Peninsula, *Pistacia mutica* is found in valleys in the western piedmont, particularly in the Bel'bek

Valley and its tributaries, and on the coast between the mouths of the Bel'bek and Kacha rivers (Figure 4.2A).

Maple in Crimea is represented by *Acer campestre* and *Acer stevenii*, neither of which is dominant in its communities. The more widespread of the two species is *A. campestre* (field maple), which is found in most of Crimea's forests, particularly in valleys. *Acer stevenii* (Steven's maple) has a more restricted distribution; it thrives mainly in stream valleys on the northern slope, where it has migrated to higher ground (Rubtsov 1978). Although endemic to Crimea, its alternative name, *Acer hyrcanum* subsp. *stevenii* Pojark, suggests a close relation to other maples in the Caucasus, Turkey and Iran.

The trees of the Betulaceae family found in the Crimean forests are *Corylus avellana*, *Alnus glutinosa*, and *Betula pendula*. An understory shrub, *Corylus avellana* (hazel) is common in forests in the mountains and in tree groves in the piedmont. *Alnus glutinosa* (sticky alder) is a cosmopolitan species that in Crimea is considered a relict element of the last glacial stage. At present, sticky alders grow mainly on north-facing slopes and along some valleys, where they are protected from radiation-generated heat and where water supply in the soil is abundant. However, pollen diagrams encompassing the Late Pleistocene and Early Holocene do not show a decline of *Alnus* pollen through the Holocene. In fact, the Buran Kaya record in the eastern piedmont shows a relative increase after 10,000 RCYBP (Gerasimenko 2007). Furthermore, *Alnus* pollen appears in relatively small amounts in the records of the Heraklean Peninsula and the Chyornaya Valley during the Holocene and up until the time of Greek colonization (Cordova and Lehman 2003).

Silver birch (*Betula pendula*), a typical boreal tree species, is found in Crimea only in a small location on the northeastern slope of the Babugan Yaila on the nearby Chamny Burun Hill at an elevation between 1,200 m and 1,300 m (Dikorastuyushchiye Rastyeniya Kryma 2012). But historical references from the nineteenth century suggest that there were more of these stands in other places in the mountains (Yena, Yena and Yena 2004b). The presence of the silver birch stand at Chamny Burun has been referred to as a glacial relict community (Kryukova et al. 1988). Paleobotanical remains (Gubonina 1985) and pollen data (Gerasimenko 2007) from deposits dated to the last glacial phase (MIS 2) in the piedmont show the presence of *Betula* possibly forming, which has been interpreted as an open woodland or a birch forest-steppe. Attempts to replant silver birch in the yailas began in the 1960s but had very little success because the constant strong winds prevented the trees from growing (Dikorastuyushchiye Rastyeniya Kryma 2012).

Other trees worthy of mention in the Crimean forests are elm, ash and linden, all of which are commonly found, although none is a dominant species. There are five native species of elm in Crimea: *Ulmus glabra* (Wych elm), *U. laevis* (white European elm), and *U. minor* (small elm) (Yena 2012). These species are commonly found in forests and forest-steppe associations, although they are never dominant. Of the ash trees, the most common in the forests of Crimea is the tall ash (*Fraxinus excelsior*), which is found mainly between 500 m and 900 m elevation. The narrow-leaf ash (*Fraxinus angustifolia*) and the raywood ash (*Fraxinus oxycarpa*) are found predominantly in the forests of the south coast (Rubtsov 1972). There are three native species of linden in Crimea: *Tilia cordata*, *T. caucasica* and *T. dasystyla* (Yena 2012). These species are found at higher forest belts in the mountains and in crevasses in the yailas (Rubtsov 1972).

Juniper in Crimea is represented by three species: *Juniperus excelsa*, *J. oxycedrus*, and *J. sabina* (Rubtsov 1972). *J. excelsa* (tall juniper) is an edificator species in co-dominance with *Pistacia mutica* and *Quercus pubescens* in some associations along the south coast. *J. oxycedrus* (prickly juniper) is a small shrub, rarely taller than 1 m, which is found on rocky calcareous substrate, particularly on the south coast and in southwestern Crimea. *J. Sabina* (Savin juniper) is found at various elevations, including the yailas.

Among other shrubs of Mediterranean origin, *Arbutus andrachne* (madrone or strawberry tree), a species of the Ericaceae family, appears as one of the typical Mediterranean floristic elements in Crimea, where it is restricted to a narrow belt between Simeiz and Alushta, although it is found in a few valleys in the west part of the piedmont, particularly near the bottom of the Uzundzha River canyon northeast of the Baydar Valley, in the southwestern part of the Crimean Mountains (Figure 4.2B).

Paliurus spina-christi, known in English as the Jerusalem thorn and in Russian as *derzhi-derevo* (the grabbing tree), is a shrub of the Rhamnaceae family, with ample distribution along the south coast and the piedmont. It is also a species of Sub-Mediterranean origin (Walter 1974) and an important component of the shiblyak communities of the South Coast (Didukh 1992). The Rosaceae shrubs are abundant in Crimea but never have a dominant role in any association. They include species in the genera *Malus*, *Pyrus*, *Crataegus* and *Rubus*.

Rhus coriaria and *Cotinus coggigrya*, both in the Anacardiaceae family, are two woody species of Sub-Mediterranean affiliation. *Rhus coriaria* (Tanner's sumac) is common in the south coast, particularly in the shiblyak community. *Cotinus coggigrya* (Eurasian smoke-tree) is a naturalized species introduced at the end of the seventeenth century (Zgurovskaya 1996) and now widespread in various communities of the south coast, the Heraklean Peninsula and the western part of the piedmont. Other Sub-Mediterranean shrubs of significance, but not dominant in their communities, are *Cornus mas* (locally known as *kizil*) and *Ligustrum vulgare* (wild privet), both of which are found at the lower elevation levels of forests.

Forbs

Forbs encompass mainly herbaceous plants that are not graminoids; forbs seldom include scrub and small shrubs. The most abundant families of forbs in Crimea are Asteraceae, Fabaceae, Cruciferae and Lamiaceae (Golubyev 1996a). Also abundant and/or widespread forbs in Crimea are species of Caryophyllaceae, Scrophulariaceae, Umbelliferae Linaceae, Plumbaginaceae, Ranunculaceae, Rosaceae and Boraginaceae. Many of the forb families, particularly Asteraceae, Cruciferae, Chenopodiaceae, Malvaceae, Ranunculaceae and Papaveraceae, are also known for their importance as weedy vegetation in farmed and/or abandoned fields (Rubtsov 1972; 1978).

The most widespread and taxonomically diverse of all the forb families is Asteraceae, which has 336 species (309 of which are native) in 86 genera (Golubyev 1996a). The species of this family occupy basically all environments available to herbaceous plants in the peninsula, including highly salinized grounds (e.g., *Saussurea salsa* and *Artemisia taurica*), stony ground in steppes (e.g., several species of *Centaurea*), and forested areas (e.g., *Lapsana intermedia* and *Leucantheum vulgare*). Similarly diverse and widespread is the Fabaceae family, which has 235 native and 11 exotic species in 37 genera (Golubyev 1996a) in a number of life forms—forbs, shrubs, perennials and annuals (Rubtsov 1972). Abundant and highly diverse in the Fabaceae are the genera *Trifolium*, *Vicia* and *Astragalus* (Dzyens-Litovskaya 1970; Rubtsov 1978).

The Cruciferae (Brassicaceae) family is also abundant, with 146 native and 12 exotic species in 60 genera (Golubyev 1996a), occupying large tracts of open land, steppe, forest-steppe, and openings in the forest, as well as disturbed and fallow fields. The most common is the genus *Sinapis*, or the mustard, which is found in abundance in fallow fields. The Lamiaceae family, with 122 native and 13 exotics in 33 genera (Golubyev 1996a), constitutes most of the aromatic forbs typical of the petrophytic steppe and the *phrygana* or *tomillares* associations in the yailas and some steppes. They include an enormous number of species in the genera *Thymus*, *Salvia*, and *Mentha*. Because they are mostly unpalatable to livestock, they are common and abundant in overgrazed lands (Cordova and Lehman 2003). Very important in the petrophytic and meadow steppes is Lilliaceae, particularly the species in the genera *Asphodelus*, *Asphodelina*, *Crocus* and *Colchicum*. They not only include a number of endemics but also a large number of rare and endangered species.

Graminoids

Graminoids include grasses and grass-like plants (e.g., sedges, rushes). Although these are the most common plants in the steppe vegetation, their occurrence and ecological role in other vegetation zones of Crimea is also important. The Poaceae (formerly known Gramineae) or grass family is the second most taxonomically diverse family in Crimea after the Ateraceae, with 228 species in 86 genera (Yena 2012). Consequently, the Poaceae is highly diverse, with adaptations to a variety of environmental conditions (e.g., salinity, shallow soils, wind, extreme cold, ground moisture, fire, and human disturbance). Grasses include various life forms (perennial, annual, tufted, rhizomatous, aquatic, etc.) and various degrees of palatability for herbivores.

In the general climatic sense, grasses are classified according to their photosynthetic pathway into C_3 and C_4, or cool season and warm season respectively. C_3 grasses are more demanding in terms of water and thrive at air temperatures between 12°C and 22°C. Their growth greatly declines at temperatures greater than 30°C, at which evaporation rates increase considerably. The C_4 grasses, on the other hand, are better adapted to higher temperatures, normally above 21°C (Winslow, Hunt Jr, and Piper 2003), which is why they dominate in tropical areas.

Given their temperature tolerances, the most common grasses in the Crimean Peninsula are those with a C_3 photosynthetic pathway. The most common genera of grasses (all C_3) in Crimea are *Festuca*, *Stipa*, *Poa* and *Agropyron* (Rubtsov 1972; Prokudin et al. 1977). On the south coast, the group of grasses of Mediterranean affiliation known as goat grasses (*Aegilops*, *Hordeum*, *Taenaterium*) is very common (Prokudin et al. 1977).

The C_4 grasses are more likely to develop under the heat of the summer, as long as water is available, or under special conditions provided by disturbance. The native C_4 grasses in Crimea are *Botriochloa ischaemum*, found in the steppe and at many locations on the south coast; *Aleuropus littoralis*, found in the salinized soils around the Sivash and saline lakes; and *Setaria viridis*, *S. glauca* and *Paspalum paspaloides*, which are more limited in distribution (Prokudin et al. 1977). Interestingly, the majority of the C_4 in Crimea are non-native species, most of which are introduced or weed grasses, as is the case of *Cynodon dactylon* (Bermuda grass), a common grass in anthropicized environments in the peninsula.

Some aquatic grasses, such as *Arundo donax* (reed grass) of the Arundinoideae family (a C_3 grass family), thrive in wetland areas and by the side of streams along with graminoids such as *Typha* (cattail) and *Juncus* (bulrush). The Cyperaceae family (the sedges), another highly diverse family of graminoids, comprises 66 species in 13 genera (Yena 2012). Of these, *Carex* is the most diverse genus, with 39 species, which makes it the most diverse genus in Crimea (Yena 2012). Sedges are found in aquatic environments and mesic environments such as meadows. Among others, the most common genera include *Cyperus*, *Cladium*, *Eleocharis*, *Scirpus* and *Schoenus*.

Adventive flora

An interesting aspect that is often omitted in studies of Crimean vegetation is the role of introduced plant species. The introduction of plants in the prehistoric past remains unclear but it probably started on a large scale during the time of Greek colonization, in the First Millennium BC. The best record of introductions exists for recent centuries, particularly since the foundation of Nikitsky Botanical Gardens, whose original purpose was to acclimatize subtropical vegetation to the southern parts of the Russian Empire. This fact made Crimea a laboratory for the adaptation of exotic plants of economic and ornamental importance. But even accepting the Nikitsky Botanical Gardens experiments, the number of exotic plants in Crimea is astonishing. Some of the introduced species have even become naturalized and are now part of the local plant communities.

Among the adventive trees, *Juglans regia* (the common walnut) figures as one of the most important introductions in antiquity. Today it is widely cultivated, and in some cases escaped varieties have become part of the forests—many can be observed in the southwest, particularly in the valleys around the Mangup fortress. Other adventive trees include *Castanea sativa* (chestnut), a typical tree of the Euxinian forests of the southern coast of the Black Sea, and *Robinia pseudoacacia* (locust tree), a relatively recent introduction from the southeastern United States.

Cupressus sempervirens (the Mediterranean cypress) deserves a special mention among introduced trees. Although commonly planted along roads, in gardens and on the borders of orchards and vineyards, it is now becoming part of the shiblyak. The time of its introduction is not known, although it is attributed to Greek colonization (Zgurovskaya 1996). During the Stalin era, many cypresses were destroyed during a massive cull sparked by the unfounded belief that the cypress was the source of a number of diseases (Zgurovskaya 1996). However, since the debunking of this myth by scientists, the tree has repopulated most of the south coast. *Genista monspessulana* (Montpellier blook) is another evergreen Mediterranean shrub widely distributed today in Crimea. Like other introduced Mediterranean species, *G. monspessulana* is also seen in many wild forms in the shiblyak, where it is highly distinguishable by its yellow flowers.

Vegetation regions

The major vegetation regions of Crimea (Figure 4.1) are in turn subdivided into plant associations, a concept that in the local botanical literature is defined as *phyto-coenosis* (i.e., Shelyag-Sosonko, Osychnyuk and Andriyenko 1982; Didukh 1992). The names of these associations in Russian and Ukrainian normally involve common names (Shelyag-Sosonko, Osychnyuk and Andriyenko 1982) but Latin names are used in others (e.g., Didukh 1992). Additionally, many other specialized classifications have been used extensively in the literature (e.g., Kochkin 1967; Dzyens-Litovskaya 1970; Rubtsov 1978; Podgorodyetskiy 1988; Vakhrushyeva 2009). But for the benefit of the non-specialist, Rubtsov's (1978) designations of Crimea's vegetation communities are used here for the vegetation regions of Figure 4.1, which is based on the map by the Biodiversity Support Program (1999).

The steppes and forest-steppe

In Crimea the steppe and forest-steppe plant communities have been replaced by cropland and secondary pasture. The steppe vegetation alone has been reduced to 5–10 percent of its original area (Vakhrushyeva 2009). For this reason, the names of vegetation communities in vegetation maps are often preceded by the modifier "agricultural land replacing" followed by the name of the original vegetation. For example, the designation "Agricultural land replacing the *Festuca-Stipa* grassland steppe" indicates those lands originally occupied by a steppe dominated by fescue and feather grass. These original plant formations of the steppe have been reconstructed from geobotanical transects carried out before the construction of the North Crimean Canal (Figure 4.3). Patches of these communities survive only in very small plots in reserves and research stations.

The halophytic steppe occupies the northern plains, around the Sivash and along the Tarkhankut Peninsula, as well as some coastal areas of the Kerch Peninsula, and it is characterized by the dominance of salt- and drought-tolerant plant species. The factors that influence this vegetation type are climatic (low precipitation and high evapotranspiration) and hydromorphic (high water levels that bring salts near the surface) (Mikhaylov 2009a). Because of strong soil salinization, this vegetation type is the only original steppe vegetation in Crimea that has escaped plowing almost entirely.

Based on the life forms (succulent or shrubs), ground hydrology and soil salinization, three basic subtypes of halophytic steppe are distinguishable: halophytic meadow, semiarid steppe with halophytes, and psammophytic steppe (Rubstov 1978). The halophytic meadow includes grasses such as *Aeluropus littoralis* and *Puccinellia fominii*, and a number of forbs such as *Limonium meyeri*, *L. caspium*, *Plantago maritima*, *Saussurea salsa*, *Salicornia europaeae*, *Suaeda altissima*, *Salsola laricina* and *S. soda*. The semiarid steppe with halophytes occupies higher ground, where *Artemisia caucasica*, *A. lerchiana* and the highly salt-tolerant *A. taurica* (Figure 4.3) are dominant. Among grasses, the most common are *Puccinellia fominii*, *Elytrigia elongata* and *Festuca sulcata* (Vakhrushyeva 2009). The psammophytic steppe extends mainly along the coastal areas of the Azov Sea, particularly around Cape

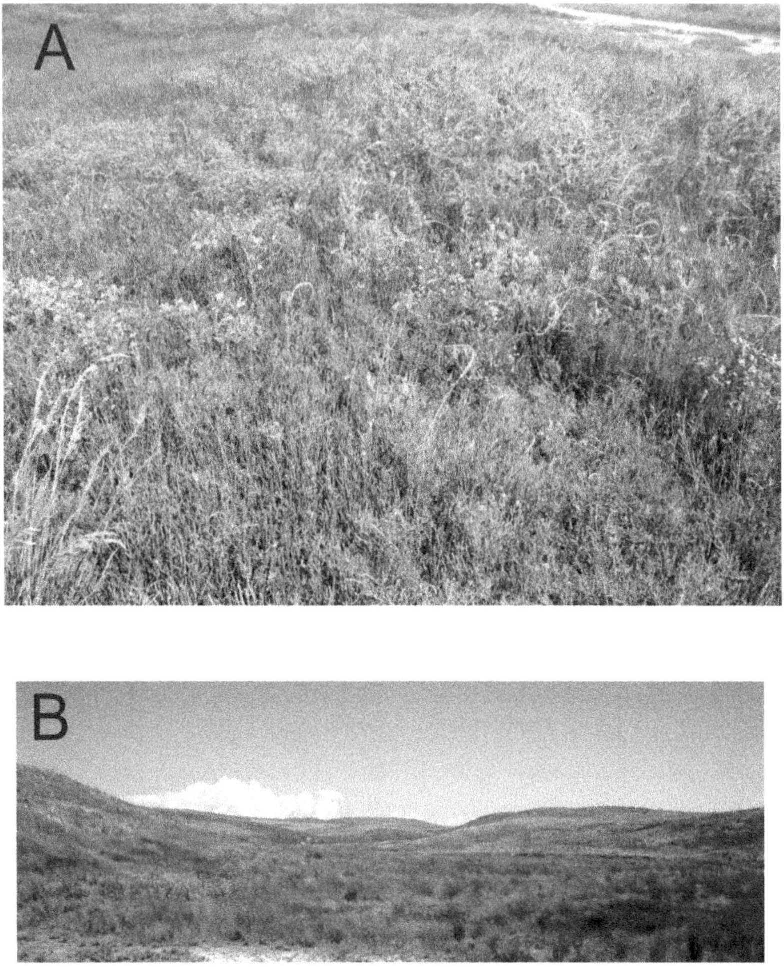

Figure 4.4: (A) The halophytic steppe, one of the few surviving types of steppe vegetation in Crimea, is dominated by *Artemisia taurica* and *A. ucrainica*; (B) True steppe dominated by *Stipa lessingiana* near Cape Tarkhan on the northwest coast of the Kerch Peninsula.

Kazantip, and along the Arabatskaya Strelka. Its common species are adapted to sand substrate, high soil salinity and dry conditions, as are the cases of *Stipa borysthenica* and *Festuca beckeri*, and forbs such as *Eryngium maritimum*, *Eurphorbia peplis*, *Asperula cimmerica* and *A. hummifusa*.

The semiarid steppe occupies an ample zone of the northern steppes, just south of the halophytic steppe (Figure 4.3). Low precipitation (250–400 mm a year) and high evapotranspiration result in a suitable environment for xerophytic plants. The typical soils of the semiarid steppe are the kastanozem, which, with the help of irrigation water, are suitable for cultivation—a reason why most of the steppe has been plowed. The original vegetation included associations dominated by small shrubs: *Artemisiataurica*, *Kochia prostata* and *Camphorosma monspeliaca*. Grasses included *Stipa ucrainica*, *S. pontica*, *Agropyron pectiniforme*, and *Festuca sulcata* among others. The composition and structure of this vegetation type is also widespread in the continental steppes across southeastern Ukraine, Russia, Kalmykia and Kazakhstan.

The true steppe used to be the most widespread of the steppe vegetation types (Figure 4.3). Its mean annual precipitation ranges between 400 mm and 500 mm a year, although values below 400 mm are common in parts of the Kerch Peninsula. The two most typical associations are those dominated by *Festuca Stipa* (fescue feather) grass steppe and the mixed-herb and *Festuca Stipa* grass steppe. The typical species of the original true steppe include, among many others, *Festuca sulcata*, *Stipa lessingiana*, *S. capillata*, *S. ucrainica* and, to a lesser degree, *Poa bulbosa* and *Agropyron pectinatus*. Among the forbs, the most common species are *Salvia nutans*, *Phlomis tuberosa*, *Medicago falcata*, *Euphorbia seguierana* and *Linum austriacum*. Because the highly productive southern chernozem soils characterize the true steppe, its original plant communities have been almost completely replaced by crops.

The petrophytic steppe is predominantly located in the Tarkhankut Peninsula and some areas in the northeastern part of the Kerch Peninsula. Although it occupies areas with precipitation similar to the true steppe, its floristic composition is different since it includes a number of species adapted to shallow, stony calcareous soils, particularly those of the Lamiaceae family. Additionally, the relatively poor retention of moisture characteristic of stony calcareous soils allows the presence of xerophytic vegetation (e.g., *Artemisia*), one reason the petrophytic steppe is often considered part of the semiarid steppe (Dzyens-Litovskaya 1970). Grasses are also important, particularly *Stipa lessingiana*.

The meadow steppe of the river valleys existed along the floodplain of the lower Salgir River and its tributaries. Despite being surrounded by territories of true steppe and semiarid steppe, the constant supply of fluvial water maintained ground moisture enough to support mesophytic forbs, of which the most important was *Beckmannia cruciformis* and several species of *Trifolium*, among others. Grasses were also important, particularly *Poa pratensis* and *Bromus inermis*.

The salinized meadow steppe is a lowland area on the southwestern coast of the Kerch Peninsula, where the high water table maintains certain salt-tolerant subaquatic species. It is geographically separated from the halophytic steppe of northern Crimea, but shares many commonalities with it, particularly with the halophytic meadow subtype. Like the halophytic steppes of the north, high salt content in the soil has helped this community escape the plow.

The meadow steppe with shrubs occupies the most humid part of the steppe, where annual precipitation is 500 mm or higher, thus forming a transition to the forest-steppe in the lower part of the piedmont. This steppe type is characterized by the mixture of forbs and grasses growing on southern chernozem soils and, due to the relatively high amounts of precipitation, mesic species are abundant. The most common of the forbs are *Trifolium repens*, *T. fragiferum*, *T. pretense*, *Vicia pannonia*, *Ononis arvensis*, *Euphorbia seguierana*, *Cichorium intybus* and *Mentha longifolia*. The dominant grasses are *Poa bulbosa*, *Poa angustifolia*, *Poa pratensis* and *Bromus ripparius* (Dzyens-Litovskaya 1970). Various species of *Festuca* and *Stipa* are also present, although not in large dominant forms as in the true steppe. Among the shrubs, the most important are *Crataegus monogyna*, *C. Taurica*, *Rosa micrantha*, *Sorbus terminalis*, *Prunus spinosa* and *Rhamnus catharticus*. Small groves of *Quercus pubescens* exist in this area, particularly towards the south.

The forest-steppe is the vegetation type that marks the transition between the open steppes of the plains and the forested areas of the piedmont and the mountains. The soils of the forest-steppe vary from southern chernozem to grey-brown forest soils and rendzinas. The vegetation communities of the forest-steppe are characterized by combinations of meadow-steppe herbs and shrubs with groves of *Quercus pubescens* (pubescent oak) and *Carpinus orientalis* (oriental hornbeam). In some areas pubescent oaks and oriental hornbeam have attained shrub size (less than 2 m) at their mature stage, a phenomenon that has been attributed to human impact, particularly the constant cutting (Vakhrushyeva 2009). The tree groves are of two types: the *dubyanik* and the *shiblyak*. The difference lies mainly in the presence of Sub-Mediterranean vegetation in the shiblyak, particularly of species such as *Pistacia mutica*, *Paliurus spina-christi* and *Ruscus ponticus* among others (Kalynychenko 2003). The dubyanik is more common in the central and eastern part as well as at higher elevations in the piedmont. The shiblyak predominates in the west, particularly in the Heraklean Peninsula and the area of the piedmont between the coast and the Kacha River valley. Other woody components in the forest-steppe include *Cotinus coggygria*, *Paliurus spina-christi* and a number of shrubs of the Rosaceae family.

The yaila steppes are completely isolated from the rest of the steppes and surrounded by forests. Plant associations vary depending on topography, orientation and substrate (Vul'f 1925; Walter 1974; Shelyag-Sosonko, Osychnyuk and Andriyenko 1982). Accordingly, there is a general trend to classify the yaila vegetation associations into meadow and petrophytic steppe types (Rubtsov 1978) (Figure 4.5), althougha spectrum of transitional herbaceous commuties (Shelyag-Sosonko, Osychnyuk and Andriyenko 1982) and wooded steppes (Vakhrushyeva 2009) does exist. In terms of the total area

Figure 4.5: Yaltinskaya Yaila. Meadow community (left), petrophytic steppe (right) and juniper woodland (background).

of the yailas, estimates indicate that between 60 percent and 80 percent is occupied by petrophytic steppes, between 20 percent and 25 percent by meadow steppe, and between 5 percent and 10 percent by wooded ecosystems (Vakhrushyeva 2009).

The petrophytic steppe is dominated by the tomillares type, which includes several species of the genus *Thymus* (thyme) in association with other forbs such as *Filipendula vulgaris*, *Medicago falcata*, *Galium verum*, *Helianthemum stevennii* and various species of the genera *Teucrium* and *Alchemilla*. Caryophyllaceae and Asteraceae have a strong representation as well, particularly *Sideritis taurica* and various species of *Centaurea*. Grasses in this type include *Stipa lithofila*, *S. tirsa* and *Alopecurus vaginatus*. The meadow type is dominated by *Carex humilis* in association with grasses and forbs. The grasses are dominated by *Stipa braunerii, Zerna riparia, Bromopsis cappadocica, Festuca pratensis, Briza elatium, Brachypodiumn pinnatum* and *Agrostis gigantea*. Forbs include a various typical meadow herbs such as *Trifolium pratense, T. repens*, *Galium verum*, *Flilipendula vulgaris*, and a number of species of *Alchemilla*, among others.

Plant associations vary also between the different yailas, particularly between the western and eastern groups (Figure 3.4). As Yevengiy Wulff (Vul'f 1925) noted, some of these differences are reflected in the dominance of *Stipa tirsae* in the eastern yailas, as opposed to the more varied set of grasses in the western yailas, particularly those of *Festuca* and *Poa*. It is possible that these differences are the result, on the one hand, of an east–west moisture gradient, which means drier conditions in the eastern Yailas (Figure 3.10A) and, on the other hand, on the fact that the eastern yailas are lower in elevation (Figure 3.5).

The broadleaf deciduous forests of the northern slopes

The forests of the northern slopes of the mountains occupy mostly the north-facing slopes of the Main Ridge and the highest slopes of the Inner Ridge (Figure 4.2C). The main forest associations are arranged vertically to form altitudinal belts (Table 4.1). The forests dominated by pubescent oak and oriental hornbeam (*Quercus pubescens* and *Carpinus orientalis*) occupy the lowermost belt, usually below 450 m, and rarely as high as 600m. Between 450 m and about 800 m, durmast oak (*Quercus petrae*) dominates on some slopes. Nonetheless, the most important forest associations at elevations higher than 500 m are those of hornbeam (*Carpinus betulus*) and common beech (*Fagus sylvatica*). English oak (*Quercus robur*) can also be found at various elevations but not as a dominant species. Of all the associations, the one that occupies the largest area is that of beech (*Fagus sylvatica*), which occupies 14 percent of all Crimea's forested areas. Its altitudinal range is usually above 450 m on both the northern and southern slopes. *Fagus sylvatica* forms associations with a number of other trees, particularly with *Carpinus betulus* and other broadleaf species such as kizilnik (*Cotoneaster tauricus*), European cornel (*Cornus mas*) and hazel (*Corylus avellana*).

Openings in the forests are often characterized by mountain meadows dominated by forbs and several species of grasses, as is the case in the Baydar Valley, where mesophytic forb elements such as *Trifolium pratense*, *T. repens*, *Filipendula vulgaris*, *Vicia cracca* and *Dictamnus gymnostilis* grow along with mesophytic grasses of the genera *Festuca*, *Melica* and *Dactylis*.

The conifer and broadleaf forests of the southern slopes

The forests of the southern slopes of the mountains are also arranged in altitudinal belts (Figure 4.2C; Table 4.1). Some broadleaf associations are similar to the ones on the northern slopes but they occur at different elevations and in association with pines.

The broadleaf forest associations include those of Mediterranean and Sub-Mediterranean woods such as *Quercus pubescens* and *Carpinus betulus*, which grow below 450 m elevation. Other broadleaf associations on the south coast include durmast oak (*Quercus petrae*), hornbeam (*Carpinus betulus*) and beech (*Fagus sylvatica*). As for the pine-dominated associations, the main formations found on the southern slope are those of Crimean pine (*Pinus pallasiana*) and Koch's pine (*Pinus sylvestris* var. *Kochiana*). Although they overlap altitudinal ranges, *Koch's pine* grows at higher elevations.

The Sub-Mediterranean shrubland and woodland

The shrublands and woodlands with Mediterranean and Sub-Mediterranean elements occupy a narrow belt along the south coast, predominantly at

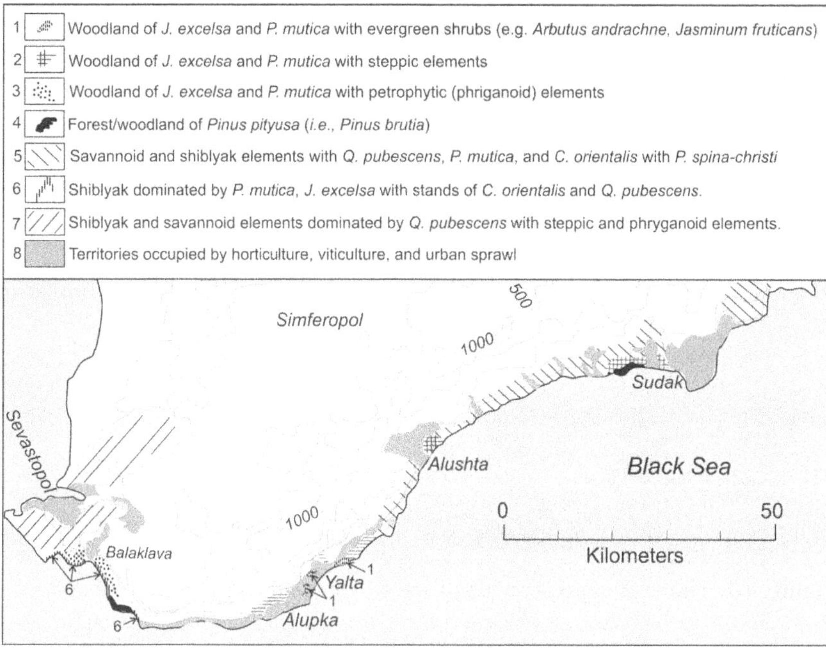

Figure 4.6: Mediterranean vegetation types. Classification based on Didukh (1992), with modifications by the author.

elevations below 250 m. Within this vegetation region, Yakov P. Didukh (1992) recognizes five vegetation associations, some of which have been further re-mapped by the author (Figure 4.6). Although the west-east precipitation gradient may influence the distribution of these associations, geological substrate and topography seem to be very important in their differentiation. Thus, those associations with phryganoid elements tend to be better established in areas of calcareous rocks, while those with shrubby and savannoid types are better established in areas of flysch, shale and volcanic rocks.

The most widespread of all the plant associations of the south coast are those dominated by tall juniper (*Juniperus excelsa*), bluntleaf pistachio (*Pistacia mutica*), shrubby forms of pubescent oak (*Quercus pubescens*) and oriental hornbeam (*Carpinus orientalis*), all of which are important elements of the shiblyak communities (Figure 4.7A–B). The shiblyaks of the Heraklean Peninsula are dense, often forming groves surrounded by phrygana or meadow steppe (Figure 4.6; formation 6). They are perhaps the transitional type between the south coast shiblyak, and the forest-steppe of the piedmont, as implied in various classifications (Garkusha 1998; Pozachyenyuk and Grishankov 1998; Kalynychenko 2003; Bondaryeva 2008).

Figure 4.7: Forms of vegetation along the south coast: (A) woodland of *Juniperus excelsa* and *Pistacia mutica* on Zolotoy Pliazh, east of Balaklava; (B) savannoid/open shiblyak with *Quercus pubescens*, *Pistacia mutica* and *Carpinus orientalis* near Morskoye; (C) open forest of *Pinus pytusae* near Cape Ayya; (D) open woodland with *Thymus* spp. phrygana.

Phrygana is also believed to be a secondary community similar to the tomillares, garrigue and macchia (Rikli 1943; Didukh 1992). Overall, it is a community dominated by petrophytic scrub (Figure 4.7D). The savannoid types include Mediterranean species of grasses, particularly of the genera *Aegilops*, *Hordeum* and *Taeniatherum*. The savannoid woodland of this region sometimes has the appearance of an open woodland of oak and pistachio (Figure 4.7B)—one reason the author refers to this association as the "open shiblyak," in contrast to the closed shiblyaks (number 6 on Figure 4.6) of some areas in the west of the south coast (Figure 4.7A).

Crimea's fauna: general aspects

The animal world of Crimea

The native fauna of the Crimean Peninsula has an interesting mix of species of diverse regions in the Palearctic realms, mirroring the distribution of flora. But unlike the flora, the composition of Crimea's faunal species has changed considerably, as animals are more likely to move in or out, or become extinct, particularly in the presence of humans. The concept of *biotope* has been introduced to place animal species and their habitats in a geographic context, often corresponding to vegetation types (Dulitskiy 2001). Accordingly, the Crimean biotopes include forest, forest-steppe, steppe and aquatic environments. The coupling of vegetation and fauna in the concept of biotope reflects primarily specialization and habitat, both of which are also important for the study of animal biodiversity and endemisms. Therefore, it is not uncommon to find that species richness and endemisms in flora often coincide with those in fauna (Figures 4.8 and 4.9).

Like flora, a number of non-native animal species are found in the peninsula. These include not only animals introduced as domesticates, but also a number of species that accompanied humans. In some cases, and particularly in the case of birds, it is difficult to draw a line between the native and non-native. Therefore, here the topic of adventives and naturalized species is circumvented and considered only for certain individual cases.

Fauna of the past

When speaking of Crimea's fauna of the past, the literature makes references to vanished and extinct species. Vanished species are those that disappeared from the peninsula but are still found elsewhere. Extinct species, however, include those that have disappeared from Crimea and elsewhere. For the study of extinct species, a distinction is often made between those that disappeared in deep geological time and those that disappeared in the Late Pleistocene and Holocene. The former are studied from an evolutionary perspective, while the latter from a historical–ecological perspective. In view of this distinction, the paleontological record of Crimea's Tertiary is full of a number of evolutionary examples, as is the case of the Camelideae and Equideae, Giraffideae, Elephantidae and Rhinocerotidae, all of which were common particularly in the Neogene (Figure 3.7). By the beginning of the Pleistocene, the Elephantidae, Rhinocerotidae and Equidae were still present. But by the beginning of the Holocene only the latter had survived.

The Eurasian elephant (*Paleoloxodon antiquus*, now *Elephas antiquus*), which along with the woolly rhinoceros (*Coelodonta antiquitatis*) was widely spread during the Early and Middle Pleistocene of Europe, is reported for the last interglacial (MIS 5e) (130,000–115,000 years BP) but became extinct during the early part of the last glacial cycle (Markova and Puzachenko 2006). *Mammuthus primigenius* (woolly mammoth) is known to have existed in the last glacial cycle (Dulitskiy 2001). Mammoth remains have been found in association with cultural remains in archaeological sites (Baryshnikov 2003; Markova and Puzachenko 2006; Patou-Mathis 2007), in paleontological localities such as Eltigen in the Kerch Peninsula (Markova and Puzachenko 2006), and karstic cavities in

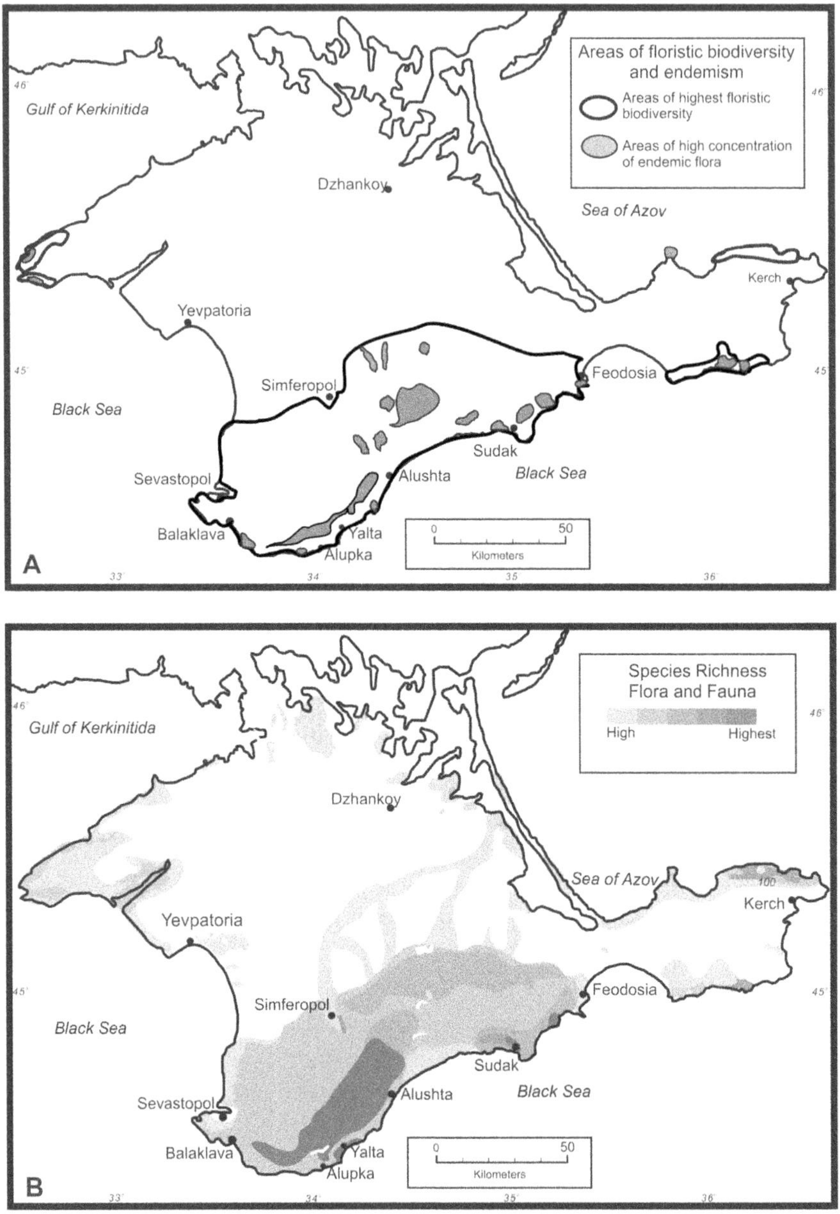

Figure 4.8: Map of biodiversity and endemisms: (A) flora; (B) flora and fauna combined. Source: Biodiversity Support Program (1999).

the Chatyr-Dag Yaila (Vremir and Ridush 2002). Dates on these localities suggest that mammoth survived probably until the beginning of the Late Glacial, at the time when the Neanderthals had practically disappeared (see Chapter 5).

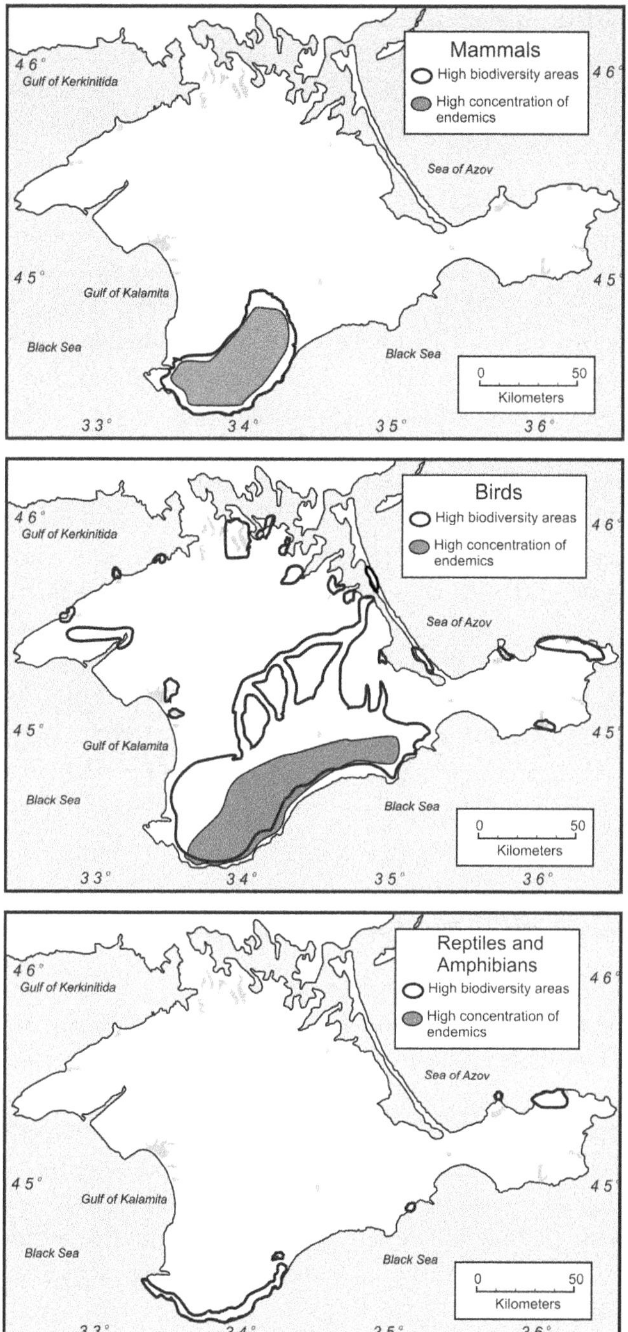

Figure 4.9: Areas of high faunal species richness and endemisms for mammals, birds, reptiles and amphibians. Source: Biodiversity Support Program (1999).

The historical–ecological examples of faunal vanishings and extinctions are more closely related to the presence of humans. The historical record has provided insight to some other species that once inhabited Crimea, particularly in the accounts of the eighteenth- and nineteenth-century naturalists, and even the early twentieth-century ones (e.g., Mokrzhyetskiy 1914). The faunal record from archaeological excavation of late prehistoric and even historic sites is also important in providing information on species that no longer inhabit Crimea. For example, the saiga (*Saiga tatarica*), an ungulate species whose distribution included Crimea during the glacial phases and even during part of the Holocene, retreated to the Kalmykian region (lower Volga) and central Asia (Dulitskiy 2001), for which the biogeographical isolation of Crimea has been proposed as one of the causes for its withdrawal in the Holocene (Vremir 2004).

Reintroductions of vanished fauna, as well as introductions of alien fauna, are widely documented. The wild boar was exterminated in the peninsula in the nineteenth century and reintroduced in 1957, and since then its populations has increased considerably (Dulitskiy 2001). The wisent bison (*Bison bonasus*), for which there is no record of its previous presence in the peninsula, was introduced in the twentieth century, although both species vanished again during World War II when the few species fell prey to hungry partisans, peasants and invading soldiers (Dulitskiy 2001). The only bison species found in prehistoric sites in Crimea is *Bison priscus*—a widespread species in Eurasia during the Late Pleistocene and which is now extinct (Dulitskiy 2001).

Faunal groups, distribution, biodiversity and endemism

Mammals

The mammals that today, or at some point in recent history, lived wild in Crimea belong to the following orders: Insectivora (insect eaters), Chiroptera (bats), Lagomorpha (rabbits and hares), Rodentia (rodents), Cetaceae (dolphins and whales), Carnivora (meat eaters), Pinnipedia (seals), Perissodactyla (horses) and Artyodactyla (two-hoofed ungulates) (Dulitskiy 2001). The greatest richness of mammal species in the peninsula is found in the mountains (Figure 4.9), where perhaps the cover of forest and the low prospects for agriculture have kept some species habitats out of reach of human destruction.

In the mountains, including the yaila, mammal fauna is represented by the Crimean red deer (*Cervus elaphus brauneri*), the fallow deer (*Cervus dama*) and the roe deer (*Capreolus capreolus capreolus*). In many forested areas, including the forest-steppe, the wild boar (*Sus scrofa ussuricus*) is also common. Wild horses and asses (genus *Equus*) have existed in the peninsula from the Miocene to historical times. Historically, the only wild horse in Crimea is the tarpan (*Equus gmelini*), a species that inhabited the steppes until the early twentieth century (Dulitskiy 2001).

Bats are a highly diverse group of mammals, of which the most common in Crimea are *Sorex aranneus araneus*, *Mustella nivalis vulgaris* and *Nyctalus leiseleri*, all of which are common in the forests, the forest-steppes and, to a lesser degree, in the steppes. Among the insectivores, *Erinaceus concolor rumanicus* is widely distributed in the forests and forest-steppes, as well as in some areas of the steppe. Rodents are also highly varied and widely distributed, occupying every biotope. Previous to the construction of the Northern Crimean Canal they were highly numerous in the steppe, but today the most common rodents in the steppes are the jerboas (*Alacctaga major* and *Dippus sagitta*), the squirrels (*Spermophilus pygnaceus brauneri*) and hamsters (*Cricetus cricetus tauricus* and *Cricetulus cricetus migratorius*).

The carnivores (order Carnivora) in Crimea include the wolf (*Canis lupus*), raccoon dog (*Nyctereutes procynoides*) and brown bear (*Ursus arctos*). Eradicated from Crimea by the beginning of the twentieth century, isolated occurrences of wolves coming from the mainland have been noted in 1952, 1960 and 1975 (Dulitskiy 2001). Many of the cats present in neighboring regions such as Asia Minor, the Balkans and the Caucasus are absent in Crimea. At present, the Crimean carnivores are basically reduced to the red fox (*Vulpes vulpes*), martens (*Martes martes* and *Martes foina rosanovi*) and weasels (*genus Mustela*).

Marine mammals in the seas around Crimea include three species of dolphins: *Dolphinus delphis ponticus* and *Tursiops truncates ponticus* in the Black Sea, and *Phocoena phocoena relicta* in the Azov Sea and some areas on the Gulf of Kerkinitida (Dulitskiy 2001). All have been considerably reduced in numbers due to predation and the action of humans, including the introduction of parasites (Kryukova et al. 1988). The only species of whale that used to visit the Black Sea was *Balaenoptera acutorostrata* (Dulitskiy 2001). The Mediterranean monk seal (*Monachus monachus*) used to be a frequent species on the Crimean coasts, but it disappeared in the early twentieth century (Dulitskiy 2001).

Reptiles and amphibians

Reptiles in Crimea belong to two orders: Squamata (snakes and lizards) and Testudines (turtles). The two Squamata suborders (Serpentes and Lacertilia) are highly diverse. Among the most common snakes, the meadow viper (*Vipera ursinii*) is the most widespread in the steppes, while the four-lined snake (*Elaphe quatuorlineata*) is the most common in the forests. Another important snake is the grass snake (*Natrix natrix*). Among the lizards, the most common are the Crimean lizard (*Podarcis tauricus*); the rock lizard (*Lacerta saxicola lindholmi*); and the sand lizards (*Lacerta agilis* and *Eremias arguta deserti*), which are commonly found on sandy beaches of the Kerch Peninsula. The only Testudine reported for Crimea is the European pond turtle (*Emys orbicularis*), which is found mainly in the north. One of the most endangered reptiles in Crimea is the legless lizard (*Ophisaurus apodus*), which is found on the south coast and in some parts of the mountains and piedmont. This species was originally found throughout the Mediterranean region and Black Sea region, where it has practically disappeared. Crimea is one of the few places where it is still found (Kryukova et al. 1988). One problem is that the legless lizard, although harmless, is often mistaken for a snake and so often killed, which is why it is one of the most endangered species in Crimea.

The south coast region has the most distinctive reptile fauna of Crimea, often presenting Mediterranean species. Furthermore, the south coast is the area of greatest richness for reptile and amphibian species (Figure 4.9). The most typical reptiles of the south coast include the Mediterranean gecko (*Tenuidachtylus kotschyi*), the Crimean gecko (*Mediodactylus kotschyi danilewskii*) and the leopard snake (*Elaphe situla*).

Knowledge of amphibians in Crimea lags behind that of reptiles, mainly because the populations of the former are smaller and scattered (Sharygin 1995). In principle, the relatively dry conditions of most of the peninsula reduce the potential for amphibian habitats. This situation, along with the destruction of habitats and pollution, has resulted in all amphibian species of Crimea being placed on the endangered species list (Biodiversity Support Program 1999).

Six species of amphibians have been documented in Crimea: *Triturus karelinii*, *Bombina bombina*, *Pelobates fuscus*, *Bufo viridis*, *Hyla arborea* and *Rana ridibunda* (Scherbak 1966). Of these, *Bufo viridis* (the European green toad) is the most common. The most rare is *Pelobates fuscus* (common spadefoot), which was thought to have vanished from Crimea in the first half of the twentieth century but which has recently been found in the Kara-Dag National Reserve (Sharygin 1995).

Bufo viridis and *Rana ratibida* (the Eurasian marsh frog) are the most resistant to disturbance, since they are found even in urban areas. At the beginning of the twentieth century, *R. ratibida* was cited as having existed in the steppe (Mokrzhyetskiy 1914), but it has recently migrated upstream to the mountains, where a number of ponds and reservoirs were constructed throughout the century (Sharygin 1995). Concentrated in the forested mountainous areas are the populations of *Triturus karelini* (southern crested newt) and *H. arborea* (the European tree frog), both of which have a strong affinity with the amphibian fauna of the Caucasus region (Scherbak 1966)

Despite the difficulties in mapping amphibian populations, the greatest species richness for reptiles and amphibians is located on the western part of the south coast, including parts of the adjacent mountains, the Kara-Dag National Reserve, and two spots on the northern shore of the Kerch Peninsula (Figure 4.9).

Birds

The total number of avian species in Crimea is more than 300 (Kostin and Dulitskiy 1998). The most common orders of birds include the Passeriformes (the house sparrow order) with 131 species, Charadriiformes (waders, gulls and auks) with 61 species, Falconiformes (birds of prey) with 31 species, and Anseriformes (ducks, geese and swans) with 28 species, together constituting 77.4 percent of all the birds of Crimea. The greatest species richness of birds in Crimea is found in the mountain forests and in the wetlands surrounding the Sivash and other small lagoonal systems (Figure 4.9). Endemisms, like other groups of fauna, are concentrated in the mountains region.

The large number of local bird species, and the numerous migratory species that spend little time in the peninsula, make it difficult to summarize all the species. Mention is therefore made of the endangered species. Of the 300 species reported for Crimea, 67 are listed in the *Ukrainian Red Data Book* (Kostin and Dulitskiy 1998). Among examples of endangered and rare birds in Crimea, I.V. Kryukova et al. (1988) list two species of pelicans (*Pelecanus onoctralus and P. crisprus*), the common shag (*Phalacrocorax aristotellis*), the black stork (*Ciconia nigra*), the common spoonbill (*Platalea leucorodia*), the Caspian tern (*Hydroprogne caspia*), the red-breasted goose (*Branta ruficolis*) and the ruddy shelduck (*Tadorna ferruginea*). The most endangered birds of prey are the osprey (*Pandion haliaetus*), the short-toed eagle (*Circateus gallicus*), two species of falcon (*Falco cherug* and *Falco peregrinus*), several eagles (*Heliaeetus albicilla*, *Aquila garax*, *A. heliaca* and *A. chrysaetos*) and the Eurasian eagle owl (*Bubo bubo*). Endangered scavenger birds include the Egyptian vulture (*Neophron percnopeterus*), the cinereous vulture (*Aegypius monachus*) and the griffon vulture (*Gyps fulvus*). In the steppes, two species of endangered bustards are *Otis tarda* and *O. tetrax*. The two most threatened species of birds in the Sivash area are the white-backed stilt (*Himantopus himantopus*) and the slender-billed curlew (*Numenius tenuirostris*).

Fishes

A distinction should be made here between freshwater and marine fishes, although a number of species in Crimea are both. The most common freshwater fishes include the common carp (*Cyprinus carpio carpio*), the blue bream (*Ballerus ballerus*) and several species of barbell (genus *Barbus*). The latter includes the endemic Crimean barbell (*Barbus tauricus*), which is an endangered species. Regarding freshwater fishes, Crimea has undergone many changes. Not only have several species been deliberately introduced but, in addition, some species from the mainland have migrated into the peninsula via the North Crimean Canal, which interconnects with the natural watercourses (Freshwater Ecoregions of the World 2013). In turn, these species have also migrated upstream to the mountains, endangering some endemics, either through predation or competition.

Of the marine freshwater fishes, the most common are the roach (*Rutilus heckeli*) and the zander (*Sander lucioperca*), which are common in the sea, in rivers and in the North Crimean Canal. Other common fish include the rainbow trout (*Oncorhynchus mykiss*), the redlip mullet (*Lisa haematocheilus*), the ziege (*Pelecus cultratus*) and various species of gobies (genus *Gobius*), of which there are about 200 species in Crimea (Klyukin and Korzhyenyevskiy 2008).

In the sea, the number of species is enormous. In pelagic environments (water bottoms), the most common are the garfish (*Belone belone*), the turbot (*Scophthalmus maximus*), the European plaice (*Pleuronectes platessa*) and the European flounder (*Platichthys flesus*). The latter can tolerate the hypersaline conditions of the Sivash (Klyukin and Korzhyenyevskiy 2008). Other fish in the surrounding seas include the goatfish (*Mullus barbatus ponticus*), the flathead mullet (*Mugil* spp.) and the sprat (*Clupeonella cultriventris*), a species in the sardine and herring family and of great economic importance. Of the sturgeon family, the Russian sturgeon (*Acipenser gueldenstaedtii*) and the beluga (*Huso huso*) are found in the seas around Crimea, where over-exploitation in combination with their slow reproductive period threaten their existence. The anchovy (*Engraulis encrasicaulis*), once abundant and economically important in the Azov and Black seas, has seen a sharp decline due to over-fishing and to predation by the invasive comb jelly (*Memiopsis leidyi*) (Chashchin 1996).

Insects and other invertebrates

Insects are the most diverse of all the faunal taxa, not only in Crimea but worldwide. The total number of insect species for Crimea is more than 12,000 (Kostin and Dulitskiy 1998). Of the 30 orders of insects, the most common in Crimea are the Coleoptera (beetles), Lepidoptera (butterflies and moths), Diptera (mosquitoes), Hymenoptera (wasps, bees, sawflies and ants) and Orthoptera (grasshoppers, crickets and locusts). Interestingly, despite the abundance and diversity of insects, a large number of species are rare and/or endangered, resulting in 31 Crimean insect species being listed in the *Ukrainian Red Data Book* (Kostin and Dulitskiy 1998).

Among insects, the banded demoiselle (*Caopteryx splendens*), the praying mantis (*Empusa fasciata*), the predatory bush cricket (*Saga pedo*), the endemic Crimean beetle (*Carabus tauricus*), the Hungarian beetle (*C. hungaricus*), the metallic ground beetle (*Calosoma sycophanta*), the great Capricorn beetle (*Cerambyx cerdo*) and the longicorn beetle (*Rosalia alpina*) are among the rare and endangered species (Kryukova et al. 1988).

Endangered moths and butterflies include *Manduca atropos*, *Hemaris tityus, Proserpinus Proserpina*, *Marumba quercus*, *Catacola sponsa*, *C. fraxini* and *Cariclea delphinii*, among others. One of the most endangered moths in Crimea is the Oleandra moth (*Deliphelia nerei*), which is a Mediterranean subtropical species with Crimea as its northernmost distribution limit. It is associated exclusively with the Sub-Mediterranean vegetation belt of the south coast. Of the Hymenoptera order (bees, wasps, sawflies and ants) the most endangered and rare species are *Scolia hirta*, *Melitturga clavicornis*, *Megachile rotundata* and *Rophitoides canus*, all of which are steppe species, and whose habitats have been destroyed or threatened by agriculture.

Arthropods (phylum Arthropoda) constitute those invertebrates characterized by an exoskeleton forming a segmented body. In Crimea they are represented by three orders: Araneae (common spiders), Soliflugae (camel spiders) and Scorpiones (scorpions). The most diverse of the three are the Araneae, with 344 species. The number of camel spider species is not known, and there is only one species of scorpion—the non-poisonous Crimean scorpion (*Euscorpius tauricus*), which is endemic to the southern coast of Crimea but which has close relatives in southern latitudes, particularly in the Mediterranean region.

Another significant invertebrate group in Crimea is the mollusks, of which there are 92 land and 50 aquatic species (Kostin and Dulitskiy 1998). One aquatic species and two land species are endemic, both of which are listed in the *Ukrainian Red Data Book*. Crustaceans are also highly diverse but the total number of species in Crimea is not known (Kostin and Dulitskiy 1998). Most widespread and best known are crabs, of which there is a high diversity. The most common are the big Black Sea lobster (*Homarus gammarus*) and the rocky crab (*Erriphia verrucosa*), both of which are the largest marine crustaceans in Crimea (Prusakov and Kozlov 1999).

CHAPTER 5

ICE AGE CRIMEA

The Pleistocene epoch is characterized by a cyclical sequence of cold phases, or glacials, separated by warm phases, or interglacials. But because glacial phases lasted longer (about 80 percent of the time), the Pleistocene is typically a cold geological epoch referred to as the Ice Age. The Pleistocene is also characterized by the emergence and spread of the hominins belonging to the genus *Homo* (e.g., *Homo erectus*, *Homo neanderthalensis*, and *Homo sapiens*).

The start date of the Pleistocene in the geological timeline has long been debated. But in recent years, by international convention its beginning has been put at around 2.58 million years ago (Figure 3.6). Its end is agreed at around 10,000 radiocarbon years ago, which in sidereal time is about 11,800 years BP (before present).

The Pleistocene has been arbitrarily divided into Early, Middle and Late Pleistocene (also known as Lower, Middle and Upper Pleistocene, respectively). But in a recent chronostratigraphic scheme, the International Commission on Stratigraphy (2013) divides the epoch into four stages: Gelasian, Calabrian, Ionian and Upper Pleistocene (Figure 3.7). However, for paleoclimatic purposes the Pleistocene is more commonly divided into Marine Isotopic Stages (MIS), which refer to periods of marked global temperature change recorded through oxygen isotopes ($\delta^{18}O$) measured in fossil microorganisms in sediments at the bottom of the oceans and directly from the oxygen in water from ice cores in Greenland and Antarctica (Figure 5.1).

Very scarce and undated information exists for the Crimean Pleistocene, except for the Upper Pleistocene (*c.*126,000–11,800 years BP), which has recently been studied in more detail in Crimea under the aegis of archaeological research in the piedmont (Chabai, Monigal and Marks 1999, 2004). The Upper Pleistocene encompasses Marine Isotopic Stages 5 to 2, which in turn correspond to the last interglacial and glacial phases. In the broad archaeological chronology of Eurasia, this time span corresponds to the Middle Paleolithic and Upper Paleolithic (Figure 5.1)

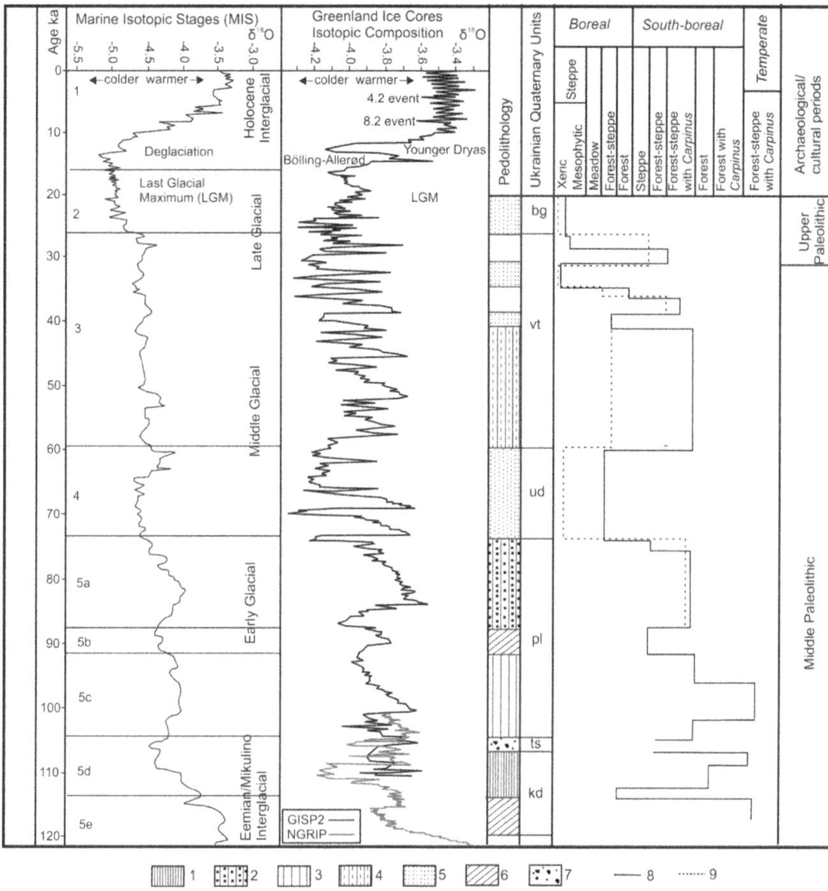

Figure 5.1: Chronostratigraphy of the Late Pleistocene in the Crimean piedmont. Soils and sediments: (1) meadow soil (mollisol); (2) rendzina and brown soil (mollic and eutric leptosol); (3) luvisol (slope derivative); (4) cambisol (or pedosediments); (5) loess (light-yellow sediment; (6) light grey loam; (7) coarse colluvial sediments. Vegetation reconstruction: (8) western Crimea; (9) eastern Crimea. Source: Gerasimenko (2007) and Cordova et al. (2011), with modifications.

Sediments and soils: witnesses of climate change

The absence of the typical glacial landforms such as cirques, moraines and other glacial features is evidence that glaciers never formed on the Crimean Peninsula during the Pleistocene. However, relicts of periglacial activity, such as rock circles, have been reported (Podgorodyetskiy 1988; Vakhrushyev and Klyukin 2001). Periglacial processes involve ice-related phenomena occurring in areas where ice is not permanent, and where soils are subjected to constant modification by freeze and thaw that results in a

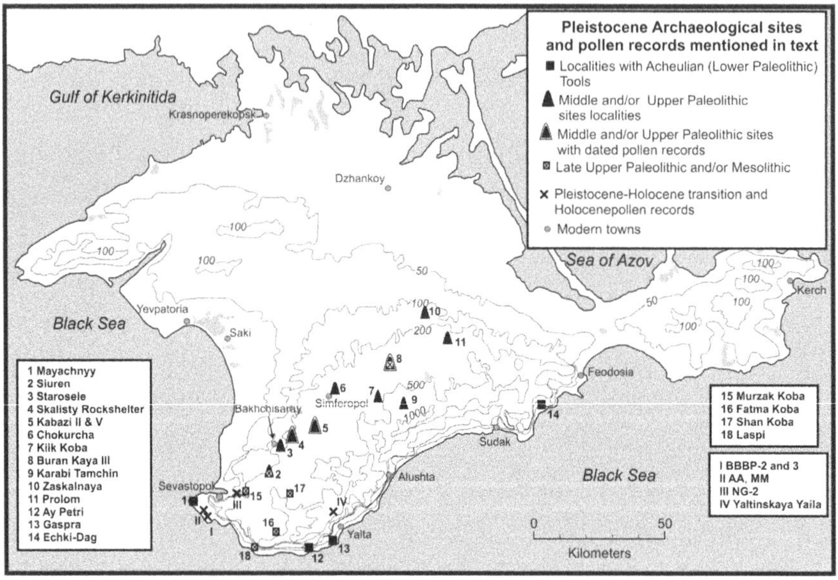

Figure 5.2: Location of the main Paleolithic sites in Crimea.

series of landforms such as rock circles and slow movements of soil downslope (i.e., solifluction).

The lack of glacier formation in the Crimean Mountains could have been the result of either, or both, of the following: temperatures were not low enough to maintain ice, and moisture was insufficient to produce snow. Indirect evidence of relatively free conditions in the Crimean Mountains during the last glaciation exists in the palynological and paleobotanical records, which show the persistence of broadleaf tree species (Gerasimenko 2007). The fauna of the peninsula during the coldest part of the glacial period was not typically boreal or arctic, but that of a steppe and forest-steppe (Markova and Puzachenko 2006). Additionally, human habitation of areas at considerable elevations in the piedmont and mountains (Figure 5.2) is indirect evidence that the mountains did not attain glacial temperatures.

Although glacial deposits are lacking in the Crimean Pleistocene, evidence of climate change exists in the sedimentary record represented by eolian deposits (i.e., loess deposits), paleosols (ancient, often buried soils), alluvial deposits and terraces, marine terraces and features under the present sea level (Figures 5.3 and 5.4). Furthermore, floral and faunal remains recovered in various paleontological and archaeological localities show extreme variations in climatic conditions throughout glacials and interglacials.

Loess and paleosol sequences

Loess is a yellow silt deposit widely distributed in mid-latitude regions of North America and Eurasia, where eolian dust was deposited during the coldest phases of the Pleistocene. In Crimea, loess deposits constitute the bulk of the Quaternary deposits attaining in some places sequences of various dozens of meters (Figure 5.3). Often these deposits are referred to as loess-like deposits or loessial clays (Podgorodyetskiy 1988). Most of the loess deposits in Crimea have been studied through deep drilling or though sediment and soil exposures along the coasts (e.g., Figure 3.13).

Loess deposits appear in stratigraphic sequences in alternation with paleosols. It is understood that loess layers correspond to glacial stages, while paleosols correspond to interglacial stages, which is why they are further correlated with MIS and their subdivisions. The loess and paleosol units form pedocomplexes, which in the Crimean Plains have been correlated with units established in the Ukrainian mainland (Vyeklich 1990; Vyeklich et al. 1993; Gozhik et al. 2001; Gerasimenko 2006). In correlation with this scheme, Natalia P. Gerasimenko (2007) classified soil development sequences in colluvial deposits into seven pedocomplex units (Figure 5.1).

Marine deposits

Marine deposits and terraces are the basis for the reconstruction of marine regressions and transgressions, sometimes combined with tectonic vertical movements.

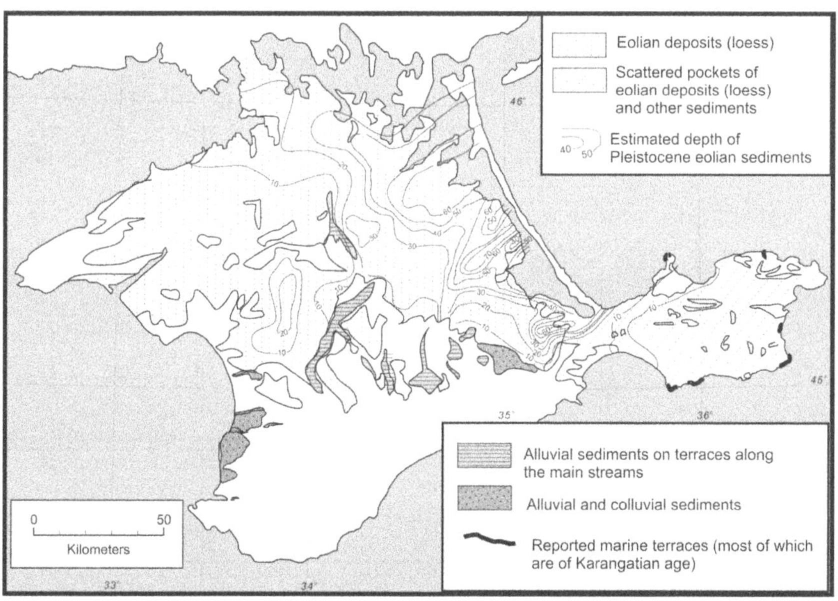

Figure 5.3: Map of Pleistocene sediments. Source: Pasynkov (2003b).

Transgressions occurred globally during interglacial periods as sea levels rose due to the melting of the glaciers. Transgression periods can be attested to by marine sediments above the modern sea level, or marine terraces, which in Crimea are located mainly on the coasts of the Kerch Peninsula and the Sudak area (Figure 5.3). The studies by Nikolay Andrusov (1912) in Sudak provided the first description of marine terraces in Crimea.

Regressions occur during glacial periods as a result of the sea-level drop caused by the locking up of the world's water in the ice caps. Regressions are more difficult to identify because their landforms are underwater but, with more recent sea sediment coring and mapping of the sea floor, some coastal features and sediments have been identified and correlated to regressions. In the Black Sea, one of the most recent major regressions is the Neo-Euxinian, which occurred in the last glacial phase, and during which a large tract of the marine shelf around Crimea was exposed (Figure 5.4).

The transgressive and regressive phases of the Pleistocene in Crimea have been studied in the coasts of the Kerch Peninsula and the Strait of Kerch (Shchyerbakov et al. 1977; Fyodorov 1978), the shelf off the southern coast

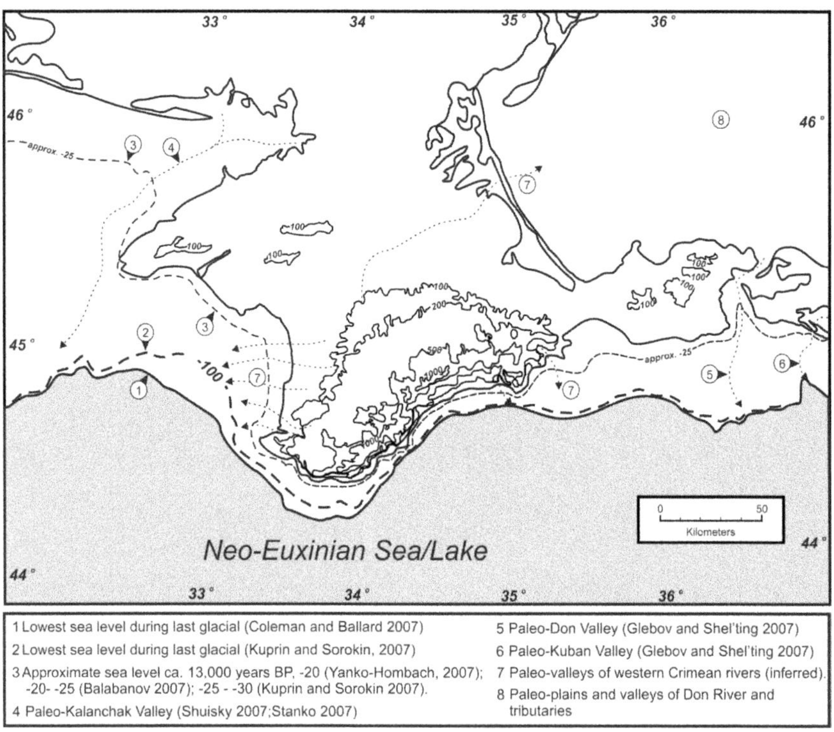

Figure 5.4: The Crimean Peninsula at glacial lowest sea level at around 20 ka BP (Last Glacial Maximum) and 13 ka BP in relation to current sea level.

(Shchyerbakov et al. 1977), and the western and northwestern shelf (Shchyerbakov et al. 1976; Gozhik and Novosyel'skiy 1989).

In the general chronology of sea-level events, the Paleo-Euxinian represents the lowering of the sea level accompanied by accumulations of eolian silts related to the Riss Glaciation (Fyodorov 1978), or MIS 6, 8, and 10 (Chepalyga 2007). The Karangat Transgression has been assigned to the Eemian interglacial or MIS 5e (Fyodorov 1978; Chepalyga 2007). The Early and Late Neo-Euxinian Regression is expressed in terrestrial deposits, in limans and on shelves, as evidence of the coldest stages (MIS 4 and 2), separated by the Surozh Transgression (occurring during MIS 3) (Chepalyga 2007).

Alluvial terraces

In correlation with the marine terraces, Nikolay Andrusov (1912) also attempted to correlate alluvial terraces in the Sudak area. Later, correlation between the Sudak terraces and the alluvial terraces of the river valleys on the west coast, was proposed leading to a general scheme of terrace seriation (Muratov 1974). This scheme is based on numbering terraces from the modern floodplain upwards, beginning with Terrace I, or the Garden Terrace (or *Sadovaya Tyerrasa* in Russian), of Terminal Pleistocene–Early Holocene age, and then older terraces in ascending order using numbers II, III, IV, V, etc.

Although no numerical ages are yet available for these terraces, relative ages have been assigned on the basis of archaeological materials in their sediments. Accordingly Terrace II has Upper Paleolithic material, while Terrace III has Late Mousterian (Late Middle Paleolithic) (Klyukin and Shchyepinskiy 1983). Later N. N. Novik (1990) proposed a chronology based on the weathering intensity of clasts in the sediments of the alluvial terraces of the Al'ma River valley. Knowing that the boundary between the Middle and Upper Paleolithic lies between 40 ka and 30 ka, the relative archaeological dates seem to be in agreement with the dates estimated by Novik (1990) based on weathering (Cordova et al. 2011). Therefore, very broadly, the ages of terraces II and III would correspond to MIS 2 and MIS 3 or earlier, respectively.

Other Pleistocene deposits and features

In a number of areas on the western coast, particularly between the Bel'bek and Al'ma rivers, and in the Heraklean Peninsula, former river beds have been filled in with Late Pleistocene and Holocene sediments at present. These ravines, often referred to as *balki* (plural: *balka*), have no running water. Instead they are occupied by meadows, and, often, cultivated land. But they contain sequences of colluvial and eolian sediments of great use for studying paleoenvironmental change in the Quaternary, as is the case of Balka Yukharina and Balka Bermana in the Heraklean Peninsula (Cordova and Lehman 2005).

Karst depressions and caves are other settings often containing Quaternary sediments. Faunal remains tend to be very rich in some of them, representing a great potential for Quaternary paleoenvironmental reconstruction (Vremir and Ridush 2002). The surfaces of the yailas are characterized by numerous depressions normally known in the karst geomorphology lexicon as uvalas, poljes and dolinas. Some of them also contain thick sequences of sediments and soils, as is the case of the Yaltinskaya Yaila pollen and sediment sequence presented and discussed in next chapter. Additionally, rock shelters in the mountains and piedmonts often have a talus of debris where sediments accumulate and where artefacts used by prehistoric dwellers are discarded, creating an archive of cultural and environmental information through millennia. In fact, many of the Pleistocene archaeological sites in Crimea are associated with rock shelter deposits (Figure 5.3). Some of these shelter sediments develop pedogenic features, useful for correlation with dated pedocomplex sequences (Figure 5.1).

Flora and fauna during the last interglacial and glacial cycles

A great deal of paleoenvironmental information on the period stretching from the last interglacial to the end of the last glacial comes from the archaeological sites in the piedmont (Figure 5.2). The most solid information regarding vegetation and soils comes from the Kabazi II, the Skalisty Rockshelter and the Buran Kaya III sites (based mainly on Gerasimenko 2004, 2007 and 2011), and is summarized in Figure 5.1. Faunal remains were also reported from the same sites, with Kabazi II having the most solid record. Additional information comes from other sites of the period, including Chokurcha I, Kabazi V, Karabi-Tamchin, Prolom II, Starosele and Zaskalnaya V (5.2).

The Eemian Interglacial (MIS 5e)

The last interglacial (MIS 5e), known as Eemian in Europe and Mikulino in Russia, was a period during which global temperatures and sea levels were similar to, or higher than, those of the present interglacial (the Holocene). The Karangat Transgression in the Black Sea left several marine terraces on the eastern Crimean coasts (Figure 5.3). In the Crimean Plains, loess deposition ceased and chernozem soils developed, constituting the Kaydaky pedocomplex unit (Vyeklich 1990).

In the stratigraphy of the Kabazi II site in the western piedmont, the first part of the interglacial is missing. The existing deposits encompass only the time after 120 ka. The deposit corresponding to this period is a light-gray loam assigned to the Kaydaky (kd) pedostratigraphic unit. Pollen from this sediment provides information on the presence of a forest-steppe characterized by oak and hornbeam woodlands alternating with meadow (mesophytic) steppes. Hornbeam (*Carpinus*), linden (*Tilia*) and elm (*Ulmus*) have an important presence in the forests of this period (Gerasimenko 2007). Overall, conditions were mesic and somewhat similar to the present.

The fauna of this period in the piedmont and mountains is dominated by forest species such as red deer (*Cervus elaphus*), forest mouse (*Apodemus flavicolus*) and forest dormouse (*Dryomys nithedula*) in combination with other typical steppe species (Markova 2000, 2005; Patou-Mathis 2005). At the site of Eltigen, on the coast facing the Strait of Kerch, the lacustine-alluvial deposits associated with a Karangat-age terrace produced a faunal assemblage dominated by steppe species such as the ground squirrel (*Spermophilus* spp.), the steppe pika (*Ochotona pusilla*) and the great jerboa (*Allactaga major*) (Markova 2005). At the Zaskalnaya site in the eastern Outer Ridge, the identified species for this period include horse (*Equus caballus*), bison (*Bison priscus*), saiga (*Saiga tatarica*) and a number of small rodents typical of the steppe (Markova 2000). No mammoth remains have been dated to this stage in Crimea but, given the relative short duration of the MIS 5e and the general scarcity of remains, it is difficult to assert that *Mammuthus primigenius* was present in the peninsula.

The interglacial-early glacial transition (MIS 5d–5a)

The early glacial, or more properly the interglacial–glacial transition, is marked by stages MIS 5d, 5c, 5b and 5a, during which considerable fluctuations in global temperature occurred as evidenced in the sea and ice $\delta^{18}O$ curves (Figure 5.1). These sharp fluctuations are reflected somehow in the paleoenvironmental record at Kabazi II, as pedogenic development and colluviation vary throughout the deposits dated to this time span. This period comprises the top half of the Kaydaky unit (kd), and the Tyasmyn (ts) and Pryluky (pl) units. The top of the Kaydaky unit is characterized by a mollisol, or dark soil formed under herbaceous vegetation similar to a chernozem, which developed during the MIS 5d. Pollen from this soil indicates an environment dominated by herbaceous vegetation, with pine trees in some areas (Gerasimenko 2007). It is during the formation of this mollisol that the first occupation takes place at the site, presumably by Neanderthals (Chabai 2005).

The Tyasmyn unit is represented at Kabazi II by a colluvial deposit with no pedogenesis. Pollen was not present in this unit (Gerasimenko 2007) but it can be assumed that this was a dry stage during which the piedmont was dominated by steppe vegetation (Figure 5.1). This unit corresponds to the coldest phase of MIS 5d in transition to MIS 5c.

The rest of the interglacial-glacial transition (MIS 5c, 5b and 5a) is characterized by the Pryluky pedostratigraphic unit, which in the plains is characterized by a kastanozem (chestnut) soil and a braunerde soil (Vyeklich et al. 1993), but which at Kabazi II is characterized by a luvisol, a light-grey loam sediment, and a rendzina that grades into a brown soil (Gerasimenko 2007). Pollen data from these soils show first the development of forest-steppe community with *Carpinus orientalis* (oriental hornbeam), which later develops into a south boreal steppe with some *Artemisia* (wormwood or sagebrush). At the top of the Pryluky unit, the pollen record shows the development of a forest-steppe in which *Betula* (birch) and *Alnus* (alder) coexist with broadleaf species such as oak and hornbeam; the strong presence of alder and Cyperaceae (sedge) pollen (Gerasimenko 2007) suggests that the locality had a cold and moist environment.

Faunal assemblages for the interglacial–glacial transition (MIS 5d to MIS 5a) are difficult to pinpoint, since data from many sites are not well dated. Overall, however, it seems that the species present during the MIS 5e interglacial remain (Markova and Puzachenko 2006). A detailed study of small mammals from Kabazi II, suggests a steady decline in forest species from MIS 5e to MIS 5b, so that at the beginning of MIS 4, the assemblage of small mammals is already dominated by steppe and meadow steppe species and a modest percentage of hydrogenous species (Markova 2005).

The Early Pleniglacial (MIS 4)

The true early glacial, often referred to as the Early Pleniglacial or Early Valdai glacial phase, corresponds to MIS 4, a time when ocean temperatures dropped significantly (Figure 5.1). The Uday (ud) pedostratigraphic unit corresponds to this period, which at Kabazi II is represented by loess (light-yellow sediment). Pollen data from this unit suggest the dominance of a boreal steppe in the western piedmont and xeric steppe in the eastern piedmont (Figure 5.1). At Kabazi II, despite the abundance of xerophytes, some hydrophytes (*Alnus* and *Cyperaceae*) thrive, suggesting the existence of local wet conditions near the site (Gerasimenko 2007).

The faunal assemblages of MIS 4 are similar to those of the previous stages (Markova 1999). However, Arctic fox (*Alopex lagopus*) has been reported from various sites including Prolom 2 in the eastern piedmont (Baryshnikov 2006). Although not properly dated, the woolly mammoth (*Mammuthus primigenius*) seems to have been present at this time in Crimea (Markova and Puzachenko 2006; Baryshnikov 2003). Other large mammal species present at this time are *Ursus speleaus* (the cave bear), *Bison priscus* (steppe bison), *Equus hydruntinus* (wild ass), *Saiga tatarica* (saiga), *Megalocerus giganteus* (Irish elk) and *Coelodonta antiquitatis* (woolly rhinoceros) (Markova and Puzachenko 2006).

The Middle Pleniglacial (MIS 3)

This Middle Pleniglacial is represented in the regional stratigraphy by the Vitaychiv soil unit (vt), which at Kabazi II has been subdivided into various soil horizons and sediments. The lowest soil corresponds to a cambisol, which is associated with the increase in broadleaf trees and the formation of a south boreal forest-steppe with hornbeam (Figure 5.1). The pollen record reveals a continuing abundance of *Pinus*, but the relatively high presence of broadleaf trees, particularly *Quercus*, *Carpinus* and *Fagus* (beech), and a relatively high frequency of Polypodiaceae spores, suggest mesic conditions under arboreal cover (Gerasimenko 2007). This moisture improvement coincides with the Surozh Transgression in the Black Sea, which in the plains causes soil salinization (Vyeklich et al. 1993).

The second half of MIS 3 saw climate deterioration as temperatures and moisture fluctuated with the alternation of stadials (cold periods) and interstadials (relatively mild periods). At Kabazi II the upper part of the Vitaychiv unit encompasses a series of colluvial and loess deposits in which pollen assemblages suggest the fluctuation between south boreal and boreal steppe and even into xeric steppe (Figure 5.1). At Buran Kaya III the pollen spectrum is dominated by a mixture of herbs, xerophytes (i.e., *Artemisia*, *Ephedra* and Chenopodiaceae) and small and variable proportions of trees (Gerasimenko 2007). In some stadials the tree pollen spectra are reduced to boreal species (*Pinus*, *Betula* and *Alnus*) and a very small proportion of broadleaf tree pollen (*Carpinus*, *Quercus* and *Tilia*) (Gerasimenko 2007).

Abundant faunal remains have been recovered from the deposits associated with MIS 3 from many sites occupied by Neanderthals as well as a number of paleontological sites (Chabai, Marks and Monigal 2004; Baryshnikov 2003; Burke 2006; Uthmeier, Ickler, and Kurbjuhn 2008). The species typical of this time include *Equus hydruntinus*, *Equus caballus*, *Saiga tatarica*, *Bison priscus*, *Mammuthus primigenius*, *Coelodonta antiquitatis*, *Rangifer tarandus* (raindeer) and *Megaloceros giganteus* (Irish elk) (Burke 2006; Markova and Puzachenko 2006; Patou-Mathis 2007). It is at the end of this period that the woolly mammoth and the woolly rhinoceros disappear from the faunal record in Crimea (Markova and Puzachenko 2006; Baryshnikov 2003). Also, at the end of MIS 3, Neanderthal occupations (marked by the Mousterian and Micocquian tools) ceased to exist in Crimea. The decline of Neanderthal populations coincides with the rapidly changing climatic conditions in Europe during MIS 3 as a result of the D-O and Heinrich events in the

Atlantic Ocean (Burroughs 2005). The $\delta^{18}O$ curve from the Greenland ice core reflects these extreme and rapid temperature fluctuations (Figure 5.1).

The Late Pleniglacial (MIS 2)

The beginning of this stage is marked by the decline of global temperatures (Figure 5.1) leading to the Last Glacial Maximum (LGM). As a consequence of this temperature decline, conditions in Crimea become colder and drier. The Bug loess unit (bg) in the plains corresponds with this cold period, which at Kabazi II is represented by a loess-like deposit mixed with colluvial gravel. Pollen assemblages from this unit indicate the existence of a north boreal steppe with a strong presence of Poaceae (grasses) and xerophytes. The arboreal pollen frequency is low (less than 20 percent for the most part) and dominated by *Pinus* (Gerasimenko 2007). Other arboreal species include *Betula*, *Alnus*, and shrubs in the Cuppresaceae, Rhamanceae, and Eleagnaceae families (Gerasimenko 2007). In the east, as usual, conditions are much drier. The Buran Kaya III diagram shows even less arboreal pollen (no more than 10 percent) with *Pinus*, *Betula*, *Alnus* and *Cupressaceae* (Gerasimenko 2007, 2011). It is possible that the array of vegetation at this time is a steppe with pine and birch in some areas and alder along streams (Gerasimenko 2011). Records of leaf impressions in travertine dated to this time show that in addition to alder, other broadleaf species such as oak, hornbeam, hazel and linden thrive in canyons (Bazhyenova 1980).

The fauna is similar to that of MIS 3, without the woolly mammoth and rhinoceros. Among the smaller mammals, the Arctic fox makes its presence again (Baryshnikov 2006). *Saiga tatarica*, *Equus ferus* and *Equus hydruntinus*, *Bison priscus*, *Ursus spelaeus*, *Megalocerus giganteus*, and *Rangifer tarandus* are still significantly present. But despite the presence of some boreal elements in the vegetation, the fauna of Crimea is not particularly that of boreal regions but of a dry steppe and forest-steppe.

The Dofinivka (Df) unit, the stratigraphic unit corresponding to the last part of the MIS 2, is characterized by weakly developed soils, still under conditions of loess deposition. Pollen records for this same period in western Crimea (Skalisty Rockshelter) show an increase in arboreal pollen (25–45 percent) and the presence of broadleaf trees (Gerasimenko 2011). In the eastern piedmont, at Buran Kaya III, a slight increase in arboreal pollen is evident, despite some gaps. Yet it seems that overall conditions at this time (around 18,000 years ago) are still of a boreal forest-steppe, as pine, birch and alder are present in the arboreal pollen spectra (Gerasimenko 2011, Figure 7). The end of the Last Glacial Maximum gives rise to the deglaciation phase, a topic that is covered in Chapter 6.

Neanderthals and modern humans in Crimea

The Paleolithic in Europe is divided into Lower, Middle and Upper Paleolithic. The Lower Paleolithic covers mainly the Lower and Middle Pleistocene, and ends roughly by the beginning of MIS 6, around 200 ka. The Middle Paleolithic encompasses the MIS 5e interglacial, the transition to glacial (MIS 5d–5a), the early glacial (MIS 4) and most of the Middle Glacial (MIS 3). The Upper Paleolithic begins towards the end of MIS 3, roughly between 38 ka and 30 ka, and extends throughout the coldest part of the last glacial stage, MIS 2, and ends during the middle part of deglaciation phase, roughly coinciding with the beginning of MIS 1 around 14,000 years ago, when a new period, the Mesolithic begins.

The Lower Paleolithic is known by the Acheulian lithic industry, which is found throughout Africa, the middle latitudes of Europe and the middle and tropical latitudes of Asia. The Acheulian tools were presumably crafted by early hominins, including *Homo erectus* and his Eurasian subspecies. Evidence of Lower Paleolithic habitation in Crimea is scarce. Traces of Acheulian lithics have been reported from various sites in the piedmont such as Bodrak, I, II and III; Zaskalnaya IX; Kara Kush I and II; Kuybishyevo, Malynovka, Prolom 3rd Terrace, and Shary I, II and III (Kolosov, Stepanchuk, and Chabai 1993). But the great majority of the Acheulian sites are located on the south coast, including Cape Mayachniy, Ay Petri, Gaspra and Echki Dag (Stepanchuk et al. 2010) (Figure 5.2). In all cases the Acheulian lithic materials are not stratified and with certainty have been reworked by geomorphological processes (Kolosov, Stepanchuk and Chabai 1993; Stepanchuk et al. 2010). Therefore, a significant archaeological record of the Lower Paleolithic is still missing in Crimea.

The better known Middle and Upper Paleolithic periods have provided more information about the hominin inhabitants of Crimea, as well as information regarding their environment and way of life. Many of their sites are stratified and often associated with caves and rock shelters, although a number of open sites occur also in areas protected by valleys. In terms of lithic industries, the Middle Paleolithic of Europe is characterized by the so-called Levalloiso-Mousterian technocomplex. The Levallois lithic technology, made since the Late Lower Paleolithic, became one of the most important features of the Middle Paleolithic throughout Europe and the Near East. The Mousterian is a style or method of tool-making that in Europe is found in association with remains of Neanderthals (*Homo neanderthalensis*), a cousin of modern humans (*Homo sapiens*), who at that time were making their way from Africa to southwest Asia. At around 30,000 years ago, when *Homo sapiens* reached Europe, the Neanderthals had practically disappeared, except for a few populations clinging to small locations in southern Spain and Portugal (Finlayson 2004). The cause of their extinction, and their possible contact with the incoming *Homo sapiensi*, has been a matter of debate among prehistorians for decades. In more recent studies there is evidence that the extinction of the Neanderthals resulted from a combination of factors, particularly problems adapting to the rapidly changing climate of Europe during MIS 3 (Burroughs 2005). The lack of adaptation to these changes is reflected in the poor nutrition found in the remains of the last populations of Neanderthals living in Europe (Finlayson 2004; Burroughs 2005).

Although their tools are found in a number of sites, Neanderthal remains in Crimea have appeared only in a few sites. The site of Kiik Koba produced the first Neanderthal remains (Bonch-Osmolovski 1925), but their exact provenance was not recorded and consequently their date is unknown. More remains were recovered from the Zaskalnaya VI site, layers III and IIIa (Kolosov, Kharitonov and Yakimov 1975). Although radiocarbon dates placed these remains between 30 ka and 39 ka, the dates should be treated with caution (see Pettit 1998). A few Neanderthal bones appear at Zaskalnaya V and a phalange in Prolom II (Kolosov, Stepanchuk and Chabai 1993; Chabai, Marks and Monigal 2004). A burial in Starosele, once deemed to be of a Neanderthal child, has recently been reinterpreted as a burial intrusion of a modern human (Marks et al. 1997). But based on associations between Neanderthal remains and Middle Paleolithic tools, it is possible to say that their presence in Crimea is apparent from around 120 ka to around 30–29 ka (Figure 5.1).

The Crimean Neanderthals made tools that fall within the Levalloiso-Mousterian and the more local technology known as the Crimean Micocquian (Chabai 2005, 2007). Their way of life was based mainly on hunting and probably scavenging. Their main prey seems to have been the wild ass (*Equus hydruntinus*) (Patou-Mathis 2005, 2007), although other faunal remains indicate a more varied source of meat (Burke 2006).

The Upper Paleolithic in Crimea signals the presence of *Homo sapiens*, known also as Cro-Magnon in Europe, and Anatomically Modern Humans (AMH) in the anthropological literature. The earliest evidence of the Upper Paleolithic in Crimea came from the Siuren I and the Buran Kaya III sites, located in the western piedmont (Figure 5.2). At the Siuren I site, Aurignacian lithics (the typical early Upper Paleolithic tool-making technique) and a modern human molar were dated between 30 ka and 28 ka (Demidenko 2008). But a direct dating of human remains at the site of Buran Kaya III puts the secure presence of Modern Humans at around 32 ka (Prat et al. 2011). These are the only two sites in Crimea, and the whole northern part of the Black Sea Region, where Middle and Upper Paleolithic occupations exist at one site.

The lithic technologies of the early Upper Paleolithic, known as Aurignacian and Gravettian were more diverse and sophisticated than the previous Levalloiso-Mousterian and Micocquian of the Neanderthals (Chabai, Marks and Monigal 2004). The technological advance of the Upper Paleolithic tool kits is clearly an adaptation to the conditions of Europe during the cold and highly variable part of the Late Glacial (Fagan 2010). The Upper Paleolithic cultures in Crimea adapted to a landscape remarkably different from the present, colder and drier climate, predominantly steppe vegetation, and a much lower coastline (Figure 5.4).

Was Crimea one of Europe's glacial *refugia*?

The biogeographical concept of *refugium* (plural: *refugia*) designates a relatively small area where populations (species) survive adverse climatic conditions or ecological change. Often the term is used in the sense of "glacial refugium," to refer to regions of southern and central Europe and North America where temperate flora and fauna retreated during the coldest phases of the Pleistocene (Burroughs 2005). The concept implies that while most of unglaciated central Europe was occupied by a steppe, many of the broadleaf temperate trees occupied protected localities in valleys, south-facing parts of the mountains, and areas of south Europe where climatic conditions were less harsh.

The lack of glacial activity on the Crimean Mountains suggests that temperatures during the cold stages of the Ice Age were relatively mild, which meant favorable conditions for tree *refugia* (Comes and Kadereit 1998; Gerasimenko 1999; Morozova and Kozharinov 2001; Leroy and Arpe 2007; Markova, Simakova and Puzanchenko 2009; Cordova et al. 2011). As an example, the pollen record for the coldest times of the last glacial period (i.e., MIS 4, the late MIS 3, and MIS 2) provides evidence of steppic vegetation with relatively low amounts of arboreal broadleaf tree pollen (mainly *Quercus* and *Carpinus*), and other shrubs (*Corylus* and Caprifoliaceae) do appear in very small amounts, particularly in Kabazi II and Skalisty Rockshelter (Gerasimenko 2007, 2011).

These pollen localities are located in valleys, which is where most likely these broadleaf species thrived. In support of this idea, records of leaf impressions in travertine roughly dating to MIS 2 in various localities in valleys and canyons in the piedmont and mountains complement the pollen record. These localities include the Big Canyon, in the upper reaches of the Bel'bek River, the Uzundha Canyon in the upper Chyornaya River basin, Krasnaya Peshchera in the upper reaches of the Salgir River, and Zhyoltaya Peshchera on the northern side of the Karabi Yaila. The species reported in the travertines are *Acer campestre*, *Alnus glutinosa*, *Carpinus betulus*, *Corylus avellana*, *Cornus australis*, *C. mas*, *Fagus sylvatica*, *Quercus petrae*, *Q. robur*, *Salix* spp. and *Tilia platyphyllos* (Bazhyenova 1980). Thus, the pollen and leaf impression records suggest that microclimatic conditions in narrow valleys favored the growth of broadleaf tree species.

Except for one impression in the Krasnaya Peshchera travertine, *Betula* is practically absent in the valleys (Bazhyenova 1980). Modest amounts of *Betula* pollen are recorded in the palynological sections of the piedmont, which suggests that birch trees were relatively widespread in this part of Crimea (Gerasimenko 2007). Pine is also absent in the leaf impression records, although it appears in relatively large amounts in the pollen records. Therefore, it is possible to picture the distribution of vegetation communities during glacial times in the mountains and piedmont of Crimea as follows: open spaces and hilltops were dominated by steppe vegetation and stands of pine and birch; valleys held broadleaf species.

The broadleaf tree glacial refugia in canyons and narrow valleys are mirrored in the modern distribution of arboreal taxa in the same localities where leaf impressions were found. It is interesting to point out that the Uzundzha Canyon today holds not only most of the trees represented in the travertine impressions dated to glacial times, but also termophilous species such as *Pistacia mutica* and *Arbutus andrachne* even at elevations around 700 m, suggesting that a warmer microclimate exists today (Figure 4.2B). These species grow mainly at low elevations along the south coast.

Another potential area for glacial *refugia* of broadleaf trees and perhaps termophilous species (i.e., Mediterranean species) could have been on the southern coast, which at that time would have been over 100 m below the present level (Yena, Yena and Yena 2004a). However, no pollen records for the Last Glacial Maximum exist anywhere for the southern slope of the mountains.

Was Crimea an island during the last interglacial?

The questions regarding the insularity of Crimea during the Karangat Transgression (MIS 5e), mentioned briefly above, is an issue that remains unsolved. The idea is that during that time the rising seas flooded the northern connection with the continent, making Crimea an island (Lazukov et al. 1981). The idea has been resurrected in association with studies of the Middle Paleolithic (Richter 2005; Chabai 2007). Indeed, it is suggested that many of the particular cultural traits of the Crimean Neanderthals—somewhat different from the rest of Europe—may be the result of such isolation (Richter 2005). However, the idea that Crimea was an island during the Karangat Transgression has been challenged on grounds of lack of stratigraphic evidence for a transgression in the Perekop-Sivash area and apparently sinking tectonics in the Sivash Graben (Blagovolin 1975), which implies that the area may have been tectonically higher at the time (Cordova et al. 2011). The latter idea implies that the connection with the mainland existed through a series of salinized steppes, lagoons, and saline lakes, not much different from the modern conditions (i.e., present interglacial conditions). These two opposing ideas, however, lack empirical evidence, which is why they must be considered hypotheses.

Jürgen Richter (2005) proposed the idea of possible primitive forms of navigation between the Crimean Karangat marine island and the mainland, which is also irrelevant in the sense that if the sea had inundated northern Crimea, the waters would have been very shallow. In fact, the depth of the Sivash is rarely above 1 m, to the point that even invading armies have crossed it on foot, a move that Neanderthals could have achieved. Given the number of bays and peninsulas in the present Sivash, it is likely that under the conditions proposed by Cordova et al. (2011), the shallow sea would have had a number of islands very close to each other. Finally, the insularity of Crimea during the Quaternary is not necessarily due to possible isolation by seas, but a biogeographical isolation, somehow linked to the concepts of the Island Biogeography Theory, which is discussed in more detail in Chapter 9.

CHAPTER 6

WARMING, RISING SEAS, AND HUMANS

The Pleistocene–Holocene transition

The Terminal Pleistocene, also known as the Deglaciation, is the period between 19,000 years BP and 11,500 years BP, roughly encompassing the last third of MIS 2 and the beginning of MIS 1. The Deglaciation is characterized by a long-term warming trend accompanied by rapid melting of ice caps and rising seas but with short cold setbacks and abrupt climatic changes (Figure 5.1, Greenland ice, $\delta^{18}O$ curve). In Crimea, this warming trend is evident in the increase in broadleaf tree pollen in the existing records, Buran Kaya III and Skalisty Rockshelter (see locations in Figure 5.2) (Gerasimenko 2004, 2011).

The pedostratigraphic sequence that corresponds to the Deglaciation in Buran Kaya III and Skalisty Rockshelter includes most of the Dofinivka unit (containing pedogenically altered loess and brown sediments) and the Prichernomorsk unit (loess-like colluvium with a weak soil) (Gerasimenko 2007). The pollen spectra associated with these sequences show an increase in broadleaf tree pollen, with two main peaks and two lows (Gerasimenko 2011). The radiocarbon dates suggest that the last peak of broadleaf tree pollen corresponds to the Bölling-Allerød phase, which in most European pollen records is expressed through the expansion of broadleaf trees that resulted from the improvement in climate (i.e., warm and moist conditions). The BBBP-2 section in the Heraklean Peninsula, which starts during the Bölling-Allerød, reflects a relatively substantial presence of broadleaf trees dominated by oak, elm and hazel (Cordova and Lehman 2005).

The Bölling-Allerød phase is followed by the Younger Dryas, which is characerized by an abrupt reduction in broadleaf trees and the establishment of steppe vegetation as a response to cooling and drying (Burroughs 2005). The Younger Dryas in the Buran Kaya III pollen record is represented by the establishment of grassy steppe vegetation and in the Skalisty Rockshelter by reduced arboreal pollen among which *Rhamnus* and *Corylus* pollen still appear (Gerasimenko 2011). It is possible that while steppization occurred in most of the piedmont, valleys held arboreal vegetation. In the BBBP-2 record, the beginning of the Younger Dryas is indicated by a decline in arboreal taxa with dominance of steppe vegetation and Rosaceae shrubs. In this same record, the development of steppic conditions is mirrored by a transition from a brown forest soil to a meadow chernozem (Cordova and Lehman 2005).

The end of the Younger Dryas is marked by the climatic amelioration in the Early Holocene, represented in the Skalisty Rockshelter pollen record as an increase in broadleaf taxa. This record, however, ends about 5,000 years BP. In the BBBP-3 pollen record, the early Holocene is characterized by the expansion of thermophilic vegetation, that is to say Sub-Mediterranean taxa such as *Quercus pubescens*-type, *Cotinus*, *Paliurus*, *Cornus* and *Verbascum* (Cordova and Lehman 2005; Cordova et al. 2011).

The most complete record for the Holocene comes from a series of sections in the Heraklean Peninsula and the Chyornaya Valley (Figure 6.1), and the Yaltinskaya Yaila pollen sequence (Figure 6.2). These records show that climatic conditions in Crimea during the Holocene fluctuate in parallel with global climatic trends, as well as with sea-level rise around the peninsula.

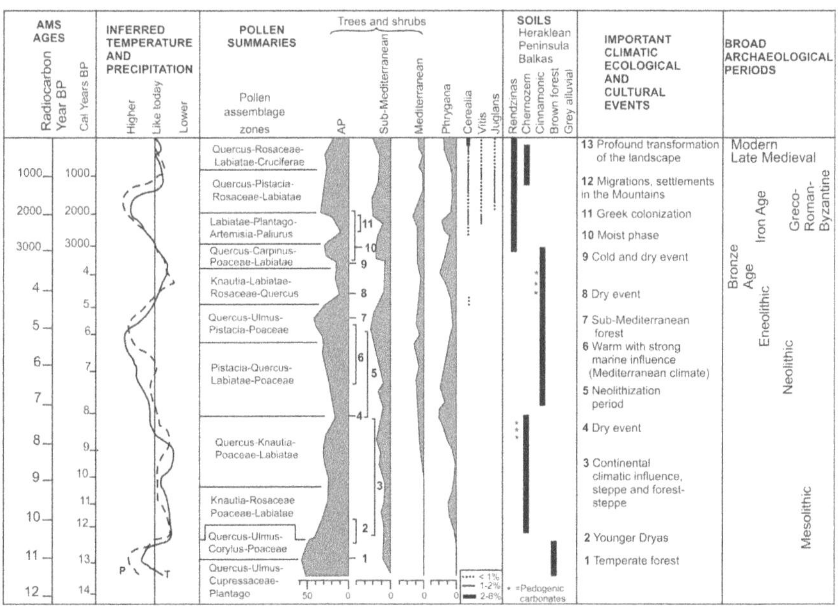

Figure 6.1: Climatic, vegetational and soil changes in southwestern Crimea during the past 14,000 years. Sources: Cordova and Lehman (2005) and Cordova et al. (2011).

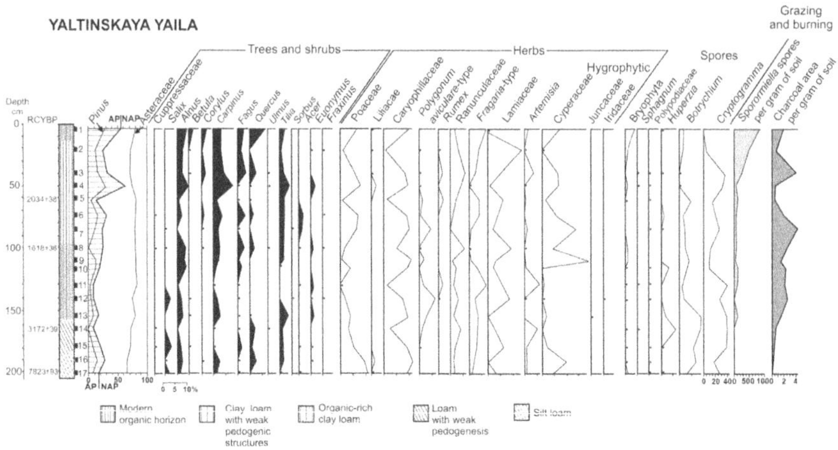

Figure 6.2: Yaltinskaya Yaila pollen diagram (AP: Arboreal pollen; NAP: Non-Arboreal pollen). Source: Cordova et al. (2011) with additional information.

Therefore, an understanding of these changes is necessary before turning to climate fluctuations and overall landscape change in the Holocene.

Postglacial rise in sea level

During the Deglaciation the coastline configuration more closely resembled the ones marked in Figure 5.4. It is not until the Early–Middle Holocene that a rapid rise in sea level around Crimea occurs. The Black Sea Flood Theory, originally proposed by William Ryan et al. (2003), suggests a rapid infilling of the Black Sea with Mediterranean waters spilling over through the Dardanelles and the Bosporus. This theory has been reviewed and discussed throughout the first decade of this century (e.g., Yanko-Hombach et al. 2007; Buynevich et al. 2011). The main consensus is that the catastrophic flooding scenario should be dismissed in view of various geological, climatic and hydrological factors. Nonetheless, the relatively rapid rise in sea level during the Middle Holocene seems to be a fact, as most sea-level curves suggest (Balabanov 2007; Martin and Yanko-Hombach 2011). Accordingly, by 6,000 years BP, the levels of the Black Sea had reached somewhere within 2 m below the present level. It is at this time that Crimea finally reaches its modern coastal configuration, except for the Sivash area, where the modern coastline does not develop until the Late Holocene.

Under this scenario, between 7,000 years BP and 6,000 years BP the shallow and low-gradient shelf around Crimea was quickly inundated by the sea. This meant that the surrounding seas began to contribute more atmospheric moisture in the form of humidity and rainfall (Cordova 2007a; Cordova et al.

2011). Therefore, a transition from a continental to a marine climate ensued. In the pollen and soil records of Crimea, this climatic transition is reflected in the increase in arboreal pollen, particularly the Sub-Mediterranean taxa, and the appearance of Mediterranean taxa (Figure 6.1). The Mediterranean taxa in the pollen diagrams include *Pistacia*, *Jasminum*, *Thymelaea hirsuta*, *Sarcopoterium*, *Asphodeline* and *Cistaceae* (Cordova and Lehman 2005; Cordova et al. 2011). Nonetheless, other climatic factors such as the temperature increase during the so-called Atlantic Phase in the Middle Holocene, should be considered (Cordova 2007).

Holocene climatic fluctuations in a variety of proxy records

Vegetation changes in Crimea during the Holocene are known through the Buran Kaya III pollen record (Gerasimneko 2004), which covers mainly the early Holocene; records obtained in several localities in southwestern Crimea, including the Chyornaya floodplain deposits (Cordova and Lehman 2005); and the Yaltinksaya Yaila (Cordova et al. 2012) (Figure 6.2). The accumulation of laminated deposits has often been cited as a proxy for reconstruction of wet and dry years. Based on data recovered by V. B. Shostakovich (1934) a somehow floating chronology of wet and dry years has been reconstructed for the period roughly after 500 AD. Using these laminated lake sediment records together with data and tree-rings Olga Solomina et al. (2005) have reconstructed a curve of wet/ dry years that at least matches some of the historical drought records for the seventeenth and eighteenth centuries. More recently, pollen data from the annual laminated sediments of Lake Saki have produced some information from around 5,380 cal. years BP (Subetto et al. 2009; Gerasimenko et al. 2011).

After the end of the Younger Dryas, conditions improved in the foothills of the mountains as steppe turned into forest-steppe around Buran Kaya III (Gerasimenko 2004). In the records of the Heraklean Peninsula (BBBP-3 and other records) the chernozem soil that developed during the Younger Dryas persisted through the early Holocene, while arboreal vegetation increased slightly, only reaching a considerable proportion (about 30 percent of the pollen spectrum) around 7,500 years BP (Cordova and Lehman 2005). Around the same time a slight peak in arboreal pollen is reached in the Yaltinskaya Yaila sequence (Figure 6.2).

During the period from 7,500 years BP to 5,000 years BP, pollen frequencies of Sub-Mediterranean and Mediterranean vegetation increase, coinciding also with the development of typical cinnamonic soils in the Heraklean Peninsula (Figure 6.1). But between 4,500 years BP and 4,000 years BP, an apparent reduction in arboreal pollen coupled with the development of pedogenic carbonates in the cinnamonic soils suggest the inception of dry climatic conditions (Cordova and Lehman 2005). Pollen and soil records from the Ukrainian mainland also show a drying trend during this period (Kremenetski 1995; Gerasimenko 1997). In the Heraklean Peninsula, strong presence of Lamiaceae (mint family) pollen during this dry phase suggests that pastoralism may have also been involved in the deterioration of vegetation (Cordova and Lehman 2005). Alternatively, the spread of Lamiaceae, typical of the petrophytic steppe, may have responded to limestone surfaces exposed by erosion. The Saki pollen record registered dry conditions during the periods 4,600–4,400, 4,200–4,100, 3,900–3,700, 3,600–3,300, with the latter being the driest (Gerasimenko et al. 2011).

In the records of the Heraklean Peninsula and the Chyornaya Valley moist conditions returned by 3,500–3,300 years BP and seem to have remained until around 2,000 years BP (approximately the beginning of the first century AD), as evidenced by the pollen record (Figure 6.1). During the same period, the Yaltinskaya Yaila pollen sequence, on the other hand, shows only a slight increase in arboreal pollen (Figure 6.2). The forest expansion that took place in the Heraklean Peninsula due to this moist phase was abruptly curbed by the establishment of ancient Greek farms (Cordova and Lehman 2003, 2005). Pollen records from the floodplain wetlands in the lower Chyornaya Valley show the appearance of cultivated grasses and an increase in weeds and phrygana elements (particularly Lamiaceae pollen) during the establishment of the Greek rural economy during the second half of the first century BC (Cordova and Lehman 2003). A pervasive dry period, which occurred before the end of the First Millennium, seems to have had a negative impact on Greek agriculture—an event discussed in more detail in Chapter 7.

The records of the NG-2 Chyornaya Valley (Cordova and Lehman 2003, 2005) and the Yaltinskaya Yaila (Figure 6.2) show a slight increase in arboreal vegetation some time between 1,500 years BP and 500 years BP, although it is not clear what this means in the context of climatic changes due to the low resolution in dates. Nonetheless, data from the Lake Saki laminated deposits provide a better picture of vegetation and climate change for the past two millennia, a subject discussed below.

Humans in the landscape

The Deglaciation in Crimea spans the late Upper Paleolithic (or Final Paleolithic in the local archaeological literature) and part of the Mesolithic (Bibikov, Stanko and Cohen 1996). The better known sites containing late Upper Paleolithic and Mesolithic have been reported mainly in the mountains and piedmont, although a considerable number of them have also been reported in the Kerch Peninsula and the plains, including the Sivash area (Bibikov, Stanko and Cohen 1998; Olenkovsky 2000). The transition from late Upper Paleolithic to Mesolithic is diffuse in terms of material culture and life ways (Cohen 2000–1), but over time the Mesolithic developed into a well-defined cultural period for which two main cultural phases are known: Shankobian and Murzakkobian.

The Shankobian culture appears around 14,000 years BP (approximately 12,800 RCYBP) and extends throughout the Pleistocene–Holocene transition. It has been studied at various sites in the piedmont and mountains, including the Shan Koba site proper, Fatma Koba, Murzak Koba, Buran Kaya and the Skalisty Rockshelter, in all of which numerous studies of lithics and faunal remains have helped archaeologists understand the ways of life of their occupants. For the most part, the faunal record suggests strong human dependence on wild sheep and goats (ovicaprids), deer, bison and/or aurochs, and wild boar, among other animals (Bibikov, Stanko and Cohen 1996). The Shankobian culture is the one that endured the cold and dry conditions of the Younger Dryas, although nothing is known as to how this climatic event affected their culture. The Swinder-type lithics, which are more common in the mainland steppes and practically unknown in Crimea, appear at the Syuren site (Cohen 2000–1). Although this appearance can be seen as a possible adaptation to steppe conditions created by the Younger Dryas, this idea is difficult to support given evidence from only a single site.

The Shankobian culture was succeeded by the Murzakkobian culture some time between 10,000 years BP and 9,000 years BP (Cohen 2000–1). The Murzakkobian includes a sub-culture known as the Kukrekian, which was relatively important in Crimea. Generally, the Murzakkobian seems to have had more geographical amplitude than the previous culture since many more sites appear not only in the piedmont but also on the south coast, including one of the best studied and dated sites: Laspi (Bibikov, Stanko and Cohen 1996; Cohen 2000–1). A number of sites are also reported in the Kerch Peninsula and many other localities in the plains. The strong presence of horses and other grazers in the faunal remains of these sites has led to the assumption that the Kerch Peninsula became a rich mesophytic steppe attractive to many grazers and hunters (Cohen 2000–1). Many sites with abundant remains of grazers have also been reported in the Sivash area (Olenkovsky 2000), where attractive conditions for grazing and a freshwater lake may have existed—with the sea level still low enough to prevent salinization and with warming conditions already in place, the Sivash area was endowed with freshwater wetlands and meadows (Podgorodyetskiy 1988). Thus, it is possible to think of a mesophytic steppe extending from the Perekop to the Kerch Peninsula, including areas now under the Azov Sea.

The Early Holocene witnessed another cultural phenomenon: Neolithization. The Neolithic period in Europe is marked by the presence of domesticated plants and animals and evidence of more sedentary ways of life, such as architecture and pottery. However, the Neolithic in Crimea is somehow different from the rest of Europe in the sense that little difference exists between the Neolithic itself and the previous cultures (the Murzakkobian and Kukrekian).

The beginning of the Crimean Neolithic is diffuse, taking place gradually sometime around or after 8,000 years BP. Thus, remains of the presumably domesticated pig (*Sus scofra*) and some pottery in archaeological sites of this period suggest the beginning of a Neolithization process (Tringham 1969; Cohen 1996). Even so, the record is still largely dominated by remains of wild animal species, suggesting high dependence on hunting (Cohen 1996). Although ovicaprids and bovines appear in the record, it seems that pig was the most common domesticated animal (Tringham 1969). Nonetheless, towards the end of the period, more evidence of other domesticated species appears (Cohen 1996).

Another important trait of the Crimean Neolithic is that many sites begin to appear at higher elevations in the mountains, including the yaila (Cohen 1996). This occurrence suggests not only the warming occurred in the Early-Middle Holocene, which allowed the exploitation of high-elevation territories, but also perhaps some sort of vertical transhumance. The Yaltinskaya Yaila section suggests that some time after 7,000 BP pedogenesis increased and that, very soon, rapid accumulation of sediments from erosion of adjacent slopes ensued. However, the pollen record from the same section does not show significant changes during this time, except for a noticeable reduction in grass and sedge pollen concomitant with the rapid accumulation of sediments (Figure 6.2).

The Eneolithic marks the beginning of a more formal Neolithization process in terms of a more sedentary life, including strong dependence on domesticates and the appearance of new and varied pottery styles. In Crimea, the Eneolithic begins around 5,000 years BP and is represented by the Kemi Oba culture, whose traits have a strong influence from the Caucasus Chalcolithic and the village cultures of mainland Ukraine (Artyemyenko 1985; Cohen 1996). The Kemi Oba culture is characterized by ritualistic traditions and social stratification, evident in the practice of dolmen-like burials and the erection of large rock monuments or menhirs (Artyemyenko 1985). One such monument exists in the Baydar Valley near Rodnivskoye (former Skelia).

The Bronze Age culture succeeds the Eneolithic between 4,500 years BP and 4,000 years BP. This early Bronze Age culture seems to have endured the climatic drying that occurred around 4,200–3,900 years BP (Figure 6.1), registered in pollen records in Crimea (Cordova and Lehman 2005) and neighboring regions (Kremenetski 1995; Gerasimenko 1997). However, it is not clear how this climatic event influenced the human population. Nonetheless, the later Bronze Age populations benefited from the climatic improvement that occurred afterwards.

The cultures associated with the Bronze Age in Crimea are the Yamnaya, Catacomb, Kamenka, Sabatinovka, Belozerka and Hallstatt. They are characterized by ritualistic burials in the so-called *kurgany* or burial mound of the steppes (Artyemyenko 1985). The Bronze Age inhabitants of the mountains and piedmont lived in small settlements and practised some sort of husbandry, most likely based on horticulture (Andreyev 1997). In the plains, nomadic pastoralism prevailed, since possibilities for agricultural development there were low. At the beginning of the First Millennium BC, the Bronze Age cultures evolved into the Iron Age culture, known in Crimea as the Kizil Koba culture, which is represented mainly by the Tauro-Scythian groups who came into contact with the Greek colonists in the First Millennium BC.

The origin of the yaila steppes

The origin of the treeless landscapes that characterize the summit of the Crimean Mountains (i.e., the yailas) has attracted the interest of various local scholars throughout the twentieth century. But with the lack of paleoenvironmental data, the explanation for the lack of trees became divided into two camps. One camp argued that the yailas were originally forested but the tree cover was gradually eliminated to create summer pastures. The other camp argued that the yailas have always had steppe vegetation due to the harsh cold and strong winds and the hydrological characteristics of its karstic terrain (Yena and Yena 2007). This situation motivated Aleksandra T. Artyushyenko and Vasiliy G. Mishnyov (1978) to examine both arguments, using pollen analysis from samples collected in a series of soil profiles in the Ai-Petrinskaya, Yaltinkskaya, Demerdzhi, Babugan and Dolgorukovskaya yailas (Figure 3.4). The sampled soil profiles were rarely deeper than 50 cm and, with no radiocarbon dates, they were deemed to have spanned the entire Holocene. Pollen data from the soil profiles showed that the frequencies of arboreal pollen throughout the profiles ranged between 10 percent and 40 percent, which was too low to support the idea that a forest cover existed on the yailas during the Holocene. Considering the possibility that most of the trees were over-represented due to the wind bringing pollen from the forests, the idea of a forest cover seemed even less possible. Based on these results, the investigators favored climatic reasons for the lack of forests on the yailas (Artyushyenko and Mishnyov 1978).

In the year 2000 the author and colleagues took a 2 m core from a depression in the western part of the Yaltinskaya Yaila to study pollen and other proxies to test further the hypothesis that the yailas have always been treeless. A first version of the pollen data (Cordova et al. 2011) shows that, for the most part, since 7,823 ± 93 RCYBP (cal. 8,524–8837 BP) the arboreal pollen percentages are low compared to percentages in modern forested areas obtained in a previous study (Cordova and Lehman 2003). Throughout the core, however, the frequency of arboreal pollen is never greater than 35 percent, except for two samples yielding 65 percent and 57 percent. However, a large percentage of the arboreal pollen sum is made up by *Pinus* (pine) pollen, a characteristic also apparent in the previous study (by Artyushyenko and Mishnyov 1978). Pine pollen is produced in abundance and usually carried very easily by the wind, to the point that large amounts of pine pollen appear even in modern samples of the Heraklean Peninsula, where there are very few pine trees (Cordova and Lehman 2003).

The potential effect of human activity was explored by measuring microscopic charcoal (<124 um) in order to evaluate tree burning, and *Sporormiella* was used to evaluate the influence of pastoralism. *Sporormiella*, a fungal spore produced in dung, is abundant in areas with large numbers of herbivores, including domesticated ones (Davis and Shafer 2006). In the sequence, microscopic charcoal concentrations begin to increase only after 3,000 years BP, but it is important to notice that the two peaks of microscopic charcoal during this period coincide with the two peaks in arboreal pollen, suggesting that burning was facilitated by an increase in wood available for fuel. *Sporormiella* concentrations seem to increase also over the past 3,000 years, with the highest concentrations only after around 1,600 years BP.

The most noticeable event in the core, involving possibly a combination of climatic and human causes, occurs after ca. 1,600 BP in samples 7 through 3, which most likely correspond to the Middle Ages (Figure 6.2). Arboreal pollen increases between samples 7 and 3, with only a slight decrease in sample 5. This sudden increase in broadleaf arboreal pollen not only coincides with erosion of the soils (the reversed radiocarbon age, suggests that) but also with the appearance of *Rumex acetosella* type, a weed strongly linked to rural disturbance by pastoralism. This phase ends in sample 2 as *Rumex acetosella* declines at the time when *Artemisia* pollen increases and sedge (Cyperaceae) and arboreal pollen frequencies decrease. This change could be an indicator of cooling and less use of the region by pastoralists. But with the lack of good resolution, this interpretation remains conjectural.

In summary, the Yaltinskaya Yaila pollen and soil sequence, and the *Sporormiella* and microscopic charcoal data suggest that during the past 8,000 years the Yaltinskaya Yaila has been for the most part a steppe and in some cases perhaps an open woodland (as it occurs in some locations today), but never a forested area. Between around 7,000 years BP and 3,000 years BP microscopic charcoal and *Sporormiella* concentrations rise only slightly, suggesting perhaps some influence of Neolithic and Eneolithic peoples, but it is not until after 3,000 years that the real human impact takes place. Although pastoralism and tree burning had an influence, they were not the decisive factor in keeping the yaila treeless.

Recent changes in sea level

Around 6,000 years BP, the Black Sea attained a level similar to its level today (the Chenomorskoye Transgression or Black Sea Transgression). During the following millennia, minor fluctuations in sea level occurred. These fluctuations have led to numerous studies that have measured the amplitude of such fluctuations. Many of these studies involved geomorphological research on archaeological monuments, including some in Crimea (e.g., Firsov 1972), and a combination of sedimentary and geomorphological work on the coasts and sea shelf (e.g., Fyodorov 1978). Later, a more comprehensive scheme resulted in the integration of data from prehistoric to medieval sites (Shilik 1997) with other data (Balabanov 2007). These schemes identified a series of transgressions and regressions in the Late Holocene, which have been named the Old Black Sea Transgression, the Phanagorian Regression, the Nymphaean Transgression, and the Medieval Regression (Figure 6.3A). This scheme, however, is open to criticism given the discrepancies between curves across the Black Sea region and the effects of tectonics (see Cordova et al. 2011). In fact, a more recent study in the area of Lake Chokrak, on the northern coast of the Kerch Peninsula, has revealed that such proposed regressions and transgressions did not really happen; instead a series of dates on the sediments of the bar separating the lake from the sea revealed a steady rise in sea level during the past 4,000 years (Kelterbaum et al. 2012). The Lake Chokrak curve is meant to allow for the offset created by tectonic movements (Figure 6.3B).

The transgression–regression curves suggest a drop in sea level during the Greek colonization of Crimea (roughly sixth to first century BC) referred to as the Phanagorian Regression. This event is followed by a period of rising sea level that caused wave erosion on abandoned Greek buildings. This rise in sea level, known as the Nymphaean transgression, coincides with the Roman–Byzantine period, and is followed by a drop in sea level, referred to as the

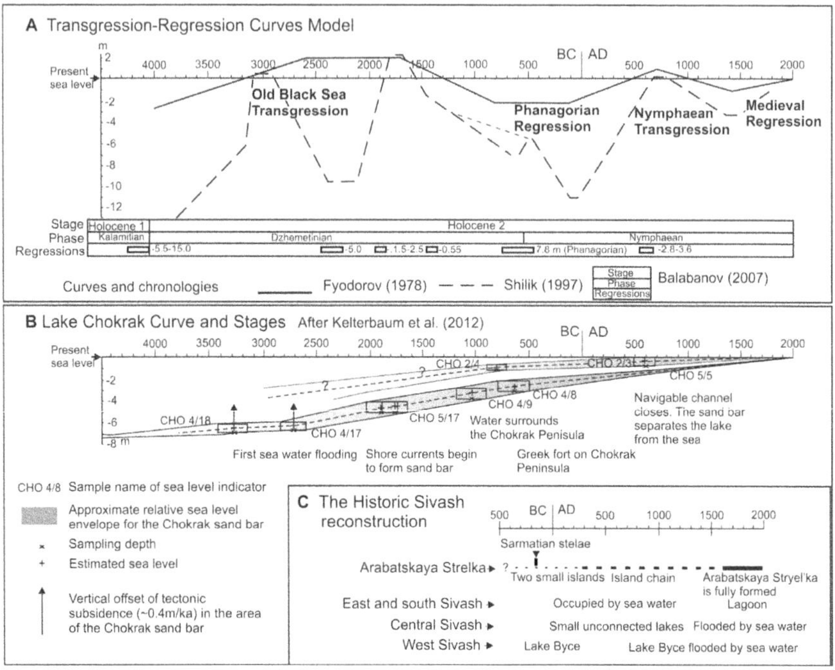

Figure 6.3: (A) Late Holocene transgression–regression model. Sea-level curves by Fyodorov (1978), Shilik (1997), and periodization by Balabanov (2007); (B) "Steady rise in sea level" model and feature changes in the Lake Chokrak sand bar (Kelterbaum et al. 2012); (C) Geomorphic events in the development of the Sivash in the Late Holocene (Cordova et al. 2011).

Medieval Regression, despite the fact that it occurs after the Middle Ages, some time in the 1400–1500s. By contrast, however, the study of Lake Chokrak reveals that the rise was continuous and without considerable fluctuations (Kelterbaum et al. 2012). Accordingly, tectonic uplift caused the ruins to rise above the rising sea level, giving the impression of subsequent regressions. In summary, rising sea level since about 4,000 years BC explains how dynamic the coast has been, and the potential changes that should be expected in coastal lakes and harbors along the coast.

The formation of the Sivash

One of the important topics regarding landscape change during the Holocene is the formation of the lagoonal system of northern Crimea—the Sivash. A number of papers published on the topic provide clues as to how the development of this interesting wetland ecosystem may have occurred, including references to historical writings and maps (i.e., Stashchuk et al.

1964; Podgorodyetskiy 1988; Yena and Yena 2008; Mikhaylov 2009a). Nonetheless, stratigraphic and geomorphic studies with chronological support have not been implemented to corroborate the historical information, which is why all the proposed scenarios are still at a hypothetical stage.

Coring of sediments below the Sivash produced stratigraphic sequences characterized by alternations of Pleistocene loess and paleosols capped by a recent sequence of lagoonal and marine sediments (Stashchuk et al. 1964). Often the distinction between lagoonal and marine sediment units is clear, as the former are grey and the latter are green (Yena and Yena 2008). Additionally, sedimentary sections exposed on the shores often show a yellow loess deposit, presumably from the last glacial phase, and a series of soil horizons, some of which are capped by a crust of salt and gypsum (Figure 3.13). This sequence suggests that the Holocene soils underwent intense salinization as the sea level and saline water table rose.

Archaeological and faunal remains in the layers showed that throughout the Upper Pleistocene, Epipaleolithic and Neolithic materials embedded in soils and deposits around the Sivash showed often dry conditions dominated by steppes, as suggested by the large number of remains of grazers found in association with the archaeological deposits (Olenkovsky 2000). Therefore, evidence of the existence of a water body like the modern Sivash is absent in the stratigraphy, suggesting that the development of the lagoonal system is Late Holocene.

Historical sources provide some insight into the formation of the Sivash, although contradictions exist amongst them. A lake east of the Perekop was reported by Strabo as Lake Sapra (Strabo, 7, 3, 4) and by Pliny the Elder as Lake Buce (Pliny 4, 12, 84). In Claudius Ptolemy's map of the Black Sea Region this presumed lake, labeled Lake Byce, is connected by a strait to Lake Maeotis (i.e., the Azov Sea). This connection is evident in Strabo's description: "It is only a part, the western part, of Lake Maeotis, for it is connected with the latter by a wide mouth" (Strabo 7, 3, 4). In Pliny the Elder's description, however, this connection is not clear, since he implies that the lake is naturally shut off from the Azov Sea and that the channel connecting the lake with the sea is artificial (Pliny, 4, 12, 84), suggesting perhaps the existence of a navigable channel. Nonetheless, Strabo reports that the lake is marshy and too shallow for navigation (Strabo 7, 3, 4).

Despite the contradictions in the historical descriptions, some of the features mentioned by the ancient authors can be located in the spatial context of the current four basins of the Sivash (Figure 1.1). Thus, Lake Sapra (or Lake Byce) most likely occupied the western Sivash basin, the gulf formed by Lake Maeotis should correspond to the modern eastern and south basins, and the strait connecting Lake Byce with Lake Maeotis was likely to be located somewhere in what is now the central basin (Stashchuk et al. 1964). With regard to the gulf presumably occupying the current eastern and southern basins, Strabo reports that "[it] contains three small islands, and also some shoals and a few reefs along the coast" (Strabo, 7, 3, 4). These "reefs" may have been a misinterpretation of shoals or barely emerged islands occupying what later became the sand bar forming the Arabatskaya Strelka (Stashchuk et al. 1964). However, the existence of islands and the incipient Arabatskaya Strelka is a matter of controversy and confusion in the ancient maps. The islands do not appear in Claudius Ptolemy's map, which was made less than a century after Strabo's description. Furthermore, a *portolan* (coastal map) dated to 1471 (in Stashchuk et al. 1964) still shows a lake with a narrow connection to the gulf, but not the islands. Also, a map by Herberstein, dated 1550, shows the archipelago of islands, but the map by Giaccobo Gaspaldo, dated 1562, does not show them. References to the area in the sixteenth and early seventeenth centuries point to the presence of the islands (Yena and Yena 2008). It is not until the late seventeenth century that the Arabatskaya Strelka is mentioned or depicted in maps (Yena and Yena 2008). In fact, the map by Mengden and Bruce, dated 1699, is perhaps the first one to show the Arabatskaya Strelka fully formed. Nonetheless, other evidence complicated this scheme. V. A. Mikhaylov (2009a) refers to findings that pose a challenge to the scenario. For example, the finding of a Sarmatian stela dated to the second to third centuries AD on the northern part of the Arabatskaya Strelka (Shults 1967) suggests that a sand bar had already emerged. However, there is the possibility that the stela was erected on one of the islands mentioned by Strabo. The main complication comes from radiocarbon dates from a lithified deposit of shell fragments in the Arabatskaya Strelka. These provided ages as early as 5,000–6,000 years BP (Balandin 1982). This dating makes little sense because of the timing of the sea-level rise reconstructed by the historical evidence (Figure 6.3). Therefore, a complex system modified by tectonics may have already been operating in the area before the historical period.

It is possible also that the presumed islands reported by Strabo in what is today the Arabatskaya Strelka may have been ephemeral, to the point that they were missed by travelers and cartographers in subsequent times. For example, they were described by Strabo towards the end of the first century BC, but not by Pliny the Elder in the following century. The shallow nature of the sea could imply that at times the islands emerged and at other times they were just sandy shoals. In fact, Strabo described the area around the islands as not navigable because it was shallow and winds could easily turn the sea into land (Strabo, 7, 4, 3).

Based on the scanty stratigraphic, geomorphological and historical information, and its relation to the transgression and regression in the late

Holocene (Figure 6.3C), the most likely scenario for the Sivash formation is the following. During the early Holocene, as the sea level was still several meters lower than today, the Sivash area was occupied by a grassy meadow steppe watered by small rivers and lakes. As the sea level rose, saline lakes and lagoons began to form. Perhaps Lake Byce was a lagoon occupying the present western basin as previously proposed (i.e., Stashchuk et al. 1964). This lagoon was connected at times to a gulf formed by the Sea of Azov in the present location of the eastern and southern basins via a strait in the location of the present central basin. It seems evident on maps that the eastern and southern basins were already separated from the Sea of Azov in the late seventeenth century (Yena and Yena 2008). The development of the sand bar at the mouth of Lake Chokrak (Kelterbaum et al. 2012) seems to have occurred sometime between 500 BC and 500 AD (Figure 6.3B). Although it is possible that the Arabatskaya Strelka was completed by the same coastal processes that formed the Lake Chokrak sand bar, stronger evidence is needed from the Arabatskaya Strelka sediments themselves.

Climate change in recent historical times

The last two millennia encompass four climatic events in the Late Holocene: the end of the Roman Climatic Optimum, the Medieval Climatic Anomaly (MCA), the Little Ice Age (LIA) and Industrial Warming (IW). The Roman Climatic Optimum, also referred to as the Graeco-Roman Climatic Optimum, roughly begins with the ascent of Rome as a power in the Italian peninsula and ends a few centuries after its decline. The Medieval Climatic Anomaly, also known as the Medieval Warm Period and less frequently as the Medieval Climatic Optimum, has been broadly defined as a relatively warm period encompassing the period 900 to 1400 AD. However, it seems that cooling spells began to put an end to the warm phase as early as the late 1200s and early 1300s (Muscheler and Fischer 2012). The LIA encompasses the time from the end of the Medieval Climatic Anomaly to the mid-1800s, after which the increase in temperatures is often paralleled with the rise in greenhouse gases from fuel burning. The truly cold nature of the LIA period is known to have encompassed at least three phases associated with low solar activity: the Spörer Minimum (1460–1550), the Maunder Minimum (1645–1715) and the Dalton Minimum (1790–1830) (Muscheler and Fischer 2012).

It is important to acknowledge that the divisions of these periods are rather arbitrary and highly variable, and that warm and cold sometimes are relative terms. Furthermore, many of these climatic events were first studied through historical documents but, as high resolution records became available, their timing and spatial extent, and their impact on natural and social systems, has

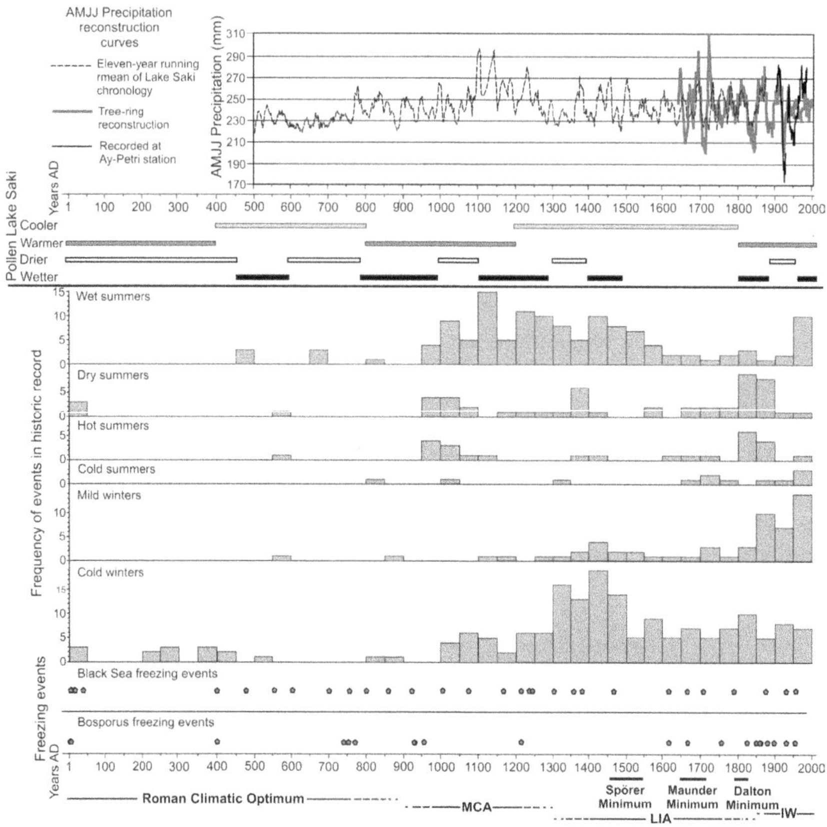

Figure 6.4: Composite of several climatic proxy data for the past two millennia. Sources: Lake Saki pollen-based reconstruction (Gerasimenko et al. 2011); historical registry for frequencies of cold winters, mild winters, hot summers, dry summers and wet summers in the Black Sea region, and freezing events on the Black Sea (Mihailescu 2004); and freezing events in the Bosporus and the Sea of Marmara (Yavuz, Akçar and Schlüchter 2007).

been reconsidered (IPCC 2007; Behringer 2010). However, evidence exists to support the view that these historic climatic events did have an impact on the society and economy of Europe (Behringer 2010). Thus, a pertinent question for the environmental history of Crimea is whether these climatic events had an impact on Crimea's environment, population and economy. To answer this question, it is important to look at historical records and cross-corroborate them with high-resolution paleoclimatic records (Figure 6.4).

The only high-resolution paleoclimatic records for the Late Holocene come from the Lake Saki's annual laminated deposits, discussed above (Shostakovich 1934; Subetto et al. 2009; Solomina et al. 2005; Gerasimenko et al. 2011). Pollen records from these laminated deposits provide information on vegetation changes (Gerasimenko et al. 2011). Additionally, for the past 300 years data from tree-rings from *Pinus sylvestris* var. *hamata* in the mountains have complemented the picture of high-resolution historical climatic oscillations in Crimea.

Historical records for climate reconstruction in the Black Sea region have been compiled from Byzantine and Ottoman documents (e.g., Grove and Conterio 1994; Mihailescu 2004; Yavuz, Akçar and Schlüchter 2007). Of these compilations, the more pertinent to Crimea, at least geographically, are those by Constantin Mihailescu (2004), from where information on dry/wet and cold/warm summers and winters can be obtained and grouped by number of events in blocks of 50 years (middle of Figure 6.4). It is important to remark that this record contains far more detailed information for the second millennium AD, which puts the Roman Climatic Optimum at a disadvantage with respect to the later periods. Finally, the historical data on freezing events in the Black Sea region and the Bosporus and Sea of Marmara (from Yavuz, Akçar and Schlüchter 2007) have also been grouped in segments of 50 years (bottom of Figure 6.4).

The records of Black Sea freezing events imply that these unusual events occurred in areas outside those parts of the sea where freezing of sea water normally occurs (i.e., the Strait of Kerch and the northwestern coast of the western Black Sea). The freezing of the Bosporus is of particular interest not only because it is at the southernmost part of the basin and the warmest part of the sea but, also, because its position at the center of the Byzantine Empire and Ottoman Empire (Constantinople) meant all freezing events must have been recorded (Yavuz, Akçar and Schlüchter 2007).

The cross-correlation of events in Figure 6.4 shows interesting patterns, but also some discrepancies, particularly concerning freezing events in the Black Sea. For example, contrary to expectations, sea-freezing events were not necessarily less frequent during the MCA and more frequent during the LIA. Nonetheless, the Bosporus data show that between 1011 and 1221, during the height of the MCA, freezing events did not occur (Figure 6.4). One possible reason for the lack of correlation between the freezing events in the Black Sea and Bosporus and the general cooling events in Europe during the LIA could be that cold frosty winters may not necessarily be linked to the same atmospheric circulation phenomena as in western Europe, where most data for the MCA and LIA have been obtained. An analysis of such events and cold winters in Istanbul in recent times fails to correlate with the North Atlantic Oscillation (NAO) phases (Yavuz, Akçar and Schlüchter 2007). An additional reason for this lack of correlation might be the strong influence of the Eurasian High during cold winters, which may interact with the effects of the NAO (Yavuz, Akçar and Schlüchter 2007). This phenomenon could also be related to the EAWR (East Atlantic–West Russia) pattern, which implies an effect from strong low pressures in the Atlantic and high pressures over the Caspian region (Oğuz, Dippner and Kaymaz 2006). Moreover, the EAWR pattern does not necessarily work in tandem with the NAO, which is why the effects of cold/dry and warm/wet winters in the Black Sea region are not particularly correlated with those over western Europe. Data from high-resolution paleoclimatic records in the Mediterranean show a see-saw effect in climatic trends between the western and eastern parts of the basin (Roberts et al. 2012). For example, records in Spain show a mostly dry climate during the MCA whereas, during the same period, records in central Anatolia show rather wet conditions (Roberts et al. 2012). It is possible that the Black Sea region is working more in parallel with the eastern Mediterranean basin. But Crimea, being farther north, may also be influenced by the effects of the EAWR pattern.

Cold winters seem to have been frequent at the beginning of the LIA around the Spörer Minimum, but not during Maunder and Dalton Minima (Figure 6.4). Mild winters seem to have been more frequent during the MCA, but they also occur with frequency in the early part of the LIA. Nonetheless, mild winters are definitely more common during the IW. Dry and hot summers seem to parallel each other, although only hot or only dry do occur. In general, however, hot and/or dry summers seem to have not been common during the MCA. On the other hand, the frequency of wet summers is relatively high during the MCA and at the end of the IW, and markedly low throughout the LIA. However, the mechanisms causing this development are not known. The NAO and EAWR patterns seem to be more related to precipitation in winter than in summer. It is possible, however, that a recurrent shift in the Westerlies may have been responsible for that pattern, perhaps by staying at that latitude during the summer.

Looking at a region farther south is useful for explaining broader climatic events. Thus, the high-resolution data gathered in the Anatolian lakes indicate a wetter MCA and drier LIA (Roberts et al. 2012), but this might be the case only for regions with winter rainfall maxima. Unfortunately, information about wet/dry winters is lacking in Constantin Mihailescu's (2004) records, although it can be deduced that the lack of cold winters during most of the MCA may have been related to wet and warmer winters. The high-resolution data gathered in the Anatolian lakes also indicate a wetter MCA and drier LIA (Roberts et al. 2012), but this might be the case for winter rainfall.

The Roman Climatic Optimum (RCO)

The so-called Roman Climatic Optimum spans the period between *c.* 100 AD and 800 AD (McCormick et al. 2012), but among scholars, the extent of its "optimum" period varies. In some cases its beginning and end are placed earlier, e.g., *c.* 300 BC to 600 AD (see Roberts et al. 2012). Although levels of moisture across Europe and the Middle East during the ROC are known as predominantly moist (McCormick 2012; Roberts et al. 2012 and references therein), major arid setbacks are evident, most notably in the period 380–540 AD (Woodbridge and Roberts 2011). Records across the Mediterranean show also that the period between 190 BC and 150 AD was a dry one in the Iberian Peninsula but not in the eastern Mediterranean (Roberts et al. 2012). Nevertheless, the sudden collapse of Greek farming in the second to first centuries BC in Crimea seems to be related to a dry phase, evident from intense soil erosion and stream incision, and somewhat represented also in pollen records for the northern Black Sea Region (Cordova et al. 2011). Some data from the Byzantine period provide evidence of cold spells during the first, fourth and seventh to ninth centuries AD (Mihailescu 2004; Yavuz, Akçar and Schlüchter 2007). The Lake Saki pollen record, however, shows that most of the Roman Climatic Optimum was dry and warm (Figure 6.4), which contrasts with the records in the Heraklean Peninsula and the Yaltinksaya Yaila (Cordova et al. 2011). Despite the cold snaps, it seems, however, that conditions may generally have been warm.

Records in many parts of the eastern Mediterranean suggest that the temperatures during the fifth and sixth centuries were relatively higher, an event that has been correlated with the first spread of the bubonic plague in the region and most notably in the Byzantine Empire, where it became known as the Justinian Plague (Roberts 2012). Named after the Byzantine emperor Justinian, who contracted the disease and survived, the plague seems to have been initiated by the global effect of several volcanic eruptions which destabilized climates in the tropics. The flea *Yersinia pestis*, the vector of the plague, originated in tropical Africa and migrated up the Nile, reaching Egypt and then the rest of the Byzantine Empire (Rosen 2008).

Warmer conditions at the time allegedly enabled the plague to spread. The high-resolution paleoclimatic records for Turkey point to increased temperatures during the time of the spread (Roberts 2011; Roberts et al. 2012). In the records compiled by Constantin Mihailescu (2004), the years between 512 and 590 AD seem to have been characterized not only by warmer conditions in the Black Sea region but also by highly variable moisture and precipitation. Under these conditions, the plague persisted until around 750 AD.

One-quarter of the population in the eastern Mediterranean region (essentially most of the Byzantine Empire) is estimated to have perished during the Justinian Plague. There are no records for the casualties in Crimea but, being part of the Byzantine Empire, it is possible that the proportion of deaths is similar to the rest of the empire. Later, in the twelfth century, the same plague returned to Europe, this time from East Asia and via the Crimean ports. This time, the so-called Black Plague claimed even more lives than the Justinian Plague.

The Medieval Climatic Anomaly (MCA)

The Medieval Climatic Anomaly (i.e., the Medieval Warm Period) extends roughly from 900 to 1300 AD, although the beginning and end dates vary among scholars. As discussed above, records for the MCA seem to indicate wetter conditions in Crimea. In particular, the laminated lake sequence in Saki Lake and tree-ring data in the mountains suggest that wet conditions prevailed at least during the period 1050–1250 AD (Solomina et al. 2012), a record that also matches the pollen data from Lake Saki (Figure 6.4).

The historical data show that summers were wetter during this period, which may have accounted for the overall increase in annual precipitation. Indirect evidence of moist and indeed optimal conditions exists in the archaeological record, which shows an unprecedented high population density in the piedmonts and mountains. Unfortunately, the two pollen records near the mountains, the Chyornaya Valley (Cordova and Lehman 2005) and the Yaltinskaya Yaila (Figure 6.2), lack the chronological resolution to show the overall vegetation response to the wet event and to the rise in population.

The Little Ice Age (LIA)

The LIA begins around 1300, although the strong cold phase does not start until the 1400s (i.e., the Spörer Minimum). The Saki Lake pollen record suggests that the period was cold, although this phase extends from around 1200 to the mid-1800s. The register of historical data by Constantin Mihailescu (2004) does show an increase in cold winters in the 1300s, 1400s, and part of the 1500s, but not later (Figure 6.5). Thus, the Maunder Minimum (1645–1715), or the coldest part of the LIA, does not seem characterized by cold winters and freezing in the Black Sea and Bosporus (Figure 6.5). Indeed, historical sources show that the Maunder Minimum was not much colder in continental European Russia than in central and western Europe (Grove and Conterio 1994). This discrepancy with western Europe may also be related to the lack of parallelism between the NAO and EAWR, as discussed above.

Industrial Warming (IW)

The end of the LIA is roughly marked in the mid-1800s, at a time when temperatures increase, in part perhaps by warming produced by industrial growth (Behringer 2010). However, no substantial quantity of data exists for Crimea before the 1900s. Temperatures throughout most of the twentieth century do not show a sharp increase (Figure 106 in Vyed' 2000). The rapid increase in temperatures does not take place until the 1970s, since when the increase has been clear and steady (Yergina 2009). The records show that during this period temperatures have increased in both winter and summer, with the latter being more pronounced, which is a pattern that is observed in most stations in Ukraine (Yergina 2009), paralleling the hike in temperatures during the past four decades in the northern hemisphere (IPCC 2007). Mean annual precipitation seems to have been higher during the first decade of the century, then slowly declines with considerably high peaks in the late 1930s, late 1940s and late 1990s, with droughts in the 1920s and 1970s (Vyed' 2000). However, unlike with temperatures, no trend in precipitation change is clear since 1900.

CHAPTER 7

FROM ANTIQUITY TO THE EIGHTEENTH CENTURY

People and environment before Greek settlement

The peoples who inhabited Crimea at the beginning of the Greek colonization (eighth to fifth centuries BC) are known through descriptions by Greek writers and from evidence in the archaeological record. Herodotus recognized three major native ethnic groups—the Scythians, the Taurians and the Cimmerians—all of which, according to linguists, spoke Indo-European languages (Andreyev 1999). Herodotus made a distinction between the Taurians, who lived in the mountainous southern part of the Crimean Peninsula, and the Scythians, who lived in the northern plains of Tauria (i.e., the Crimean Peninsula) and the plains between the Dnieper and the Don rivers (Herodotus 4, 2–103). Additionally, the Cimmerians were a group placed by Herodotus in the eastern part of Crimea, namely on the Kerch Peninsula. But by the time Greeks began to settle the region, the Cimmerians had been displaced by the Scythians (Andreyev 1999). Therefore, the two groups with whom the Greek colonists in Crimea came in contact were the Taurians and Scythians.

Historical and archaeological records suggest that the Scythians were a nomadic and semi-nomadic tribe of the steppes. Much less is known, however, about the way of life of the Taurians. The scattered archaeological record provides evidence that they were engaged in some form of horticulture and husbandry (Savelya 1996; Carter et al. 2000; Novičenkova 2008). At the Taurian sites, archaeologists have discovered goat, sheep and cattle bones as well as remains of cereals and pulses. Given their steep slopes and stony soils, the valley bottoms were the only setting where these crops could be grown. The amount of pottery together with the presence of walls at most sites is solid evidence that the Taurians had a sedentary, or at least semi-sedentary, way of life (Novičenkova 2008). The Scythians came into contact with the

Taurians, which contact ultimately ended in intermarriage and assimilation, to the point that the historical sources in the first century BC and later refer to Tauro-Scythians or Scythio-Taurians (Andreyev 1997).

Pre-Hellenistic settlement sites along the west coast provide archaeological evidence of trade with the Mediterranean peoples, most probably the Greeks. Archaeological evidence attests to the fact that early trade came from the sites of Chaika and Panskoye (Šceglov 2002) and the site of Uch Bash located in a high area south of the Chyornaya River mouth (Kravchenko 2004; Nikolaenko 2006).

Crimea in the context of ancient Greek colonization

The rapid spread of Greek settlers in the colonial enclaves of the eastern Mediterranean during the late part of the Greek Dark Ages (1200–480 BC) and through the Greek Archaic Period (800–480 BC) was the result of the unprecedented growth of the Greek population in the mainland (Martin 1996). In part, this phenomenon is attributed to the Greek system of dividing land among heirs, which in a few generations left no available land for the growing population (Hansen 1999).This prompted the Greek city states to seek land elsewhere along the Mediterranean coast.

Greek colonies in the Mediterranean began as trading points competing with the Phoenician merchants (Martin 1996). From the colonies in Asia Minor (present-day western Turkey), Greek colonists began moving towards the Sea of Marmara and the Bosporus. The Greek colonization of the Black Sea coast began in the first half of the First Millennium BC. The first Greek colonists of the Black Sea coast were the Milesians (also referred to as Ionian Greeks), who founded the cities of Olbia, Panticapaion and Istria, among others (Figure 1.4). The Megarians (also referred to as Dorian Greeks) followed suit and established their own colonies, mainly on the western half of the Black Sea Basin. It is possible that, using sea currents, they established a counterclockwise navigation route for colonization and trade between the Megaran cities of the west part of the Black Sea (Hind 1998) (Figure 1.4).

The pattern of Hellenistic settlement in Crimea and the northern Black Sea region followed the model already established in the rest of the Greek world, whereby each city-state consisted of the city (*polis*) and its agricultural hinterland (*chora*) (Bintliff 2006). In Crimea, it is possible to see the extension of the chora territories around the towns (Figure 7.1). The chora territories were divided into plots often separated by roads and stone walls, with farmhouses in between them, all of which are features clearly visible today from the air (Saprykin 1994; Carter et al. 1997; Nikolaenko 2006; Smekalova and Smekalov 2006).

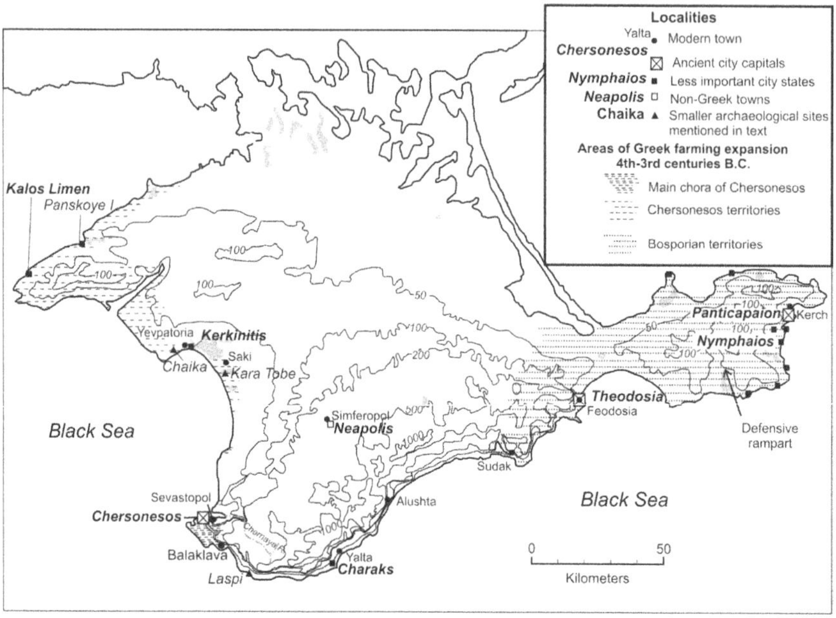

Figure 7.1: Greek cities of Crimea and their agricultural territories. Sources: Shilik (1997) and Cordova et al. (2011) with modifications.

As agriculture developed and boomed in the Black Sea coastal colonies in the fifth and fourth centuries, many of their cities became *emporia*, or centers of trade (Carter et al. 1997). Their main function was to act as intermediary points of trade between the broader Greek world and the native peoples of the hinterland. Tauric Chersonesos, for example, became an exporter of wine and fish (Sorochan, Zubar and Marchenko 2000; Carter 2003), while engaged also in trade with the Scythian tribes during times of peace.

There was a clear demarcation between the Greek farmlands and those of their native neighbors, as suggested by the series of walls and ramparts built across the Kerch Peninsula (Figure 7.1). Uneasy relations with the native populations are manifested in the fortification of some farms in the chora of Tauric Chersonesos (Saprykin 1994; Nikolaenko 1999; Sorochan, Zubar and Marchenko 2000; Carter et al. 2000). One example is Bezymyannaya, a hilltop fortress and farmstead on the southeastern border of the chora (Figure 7.2). Nonetheless, archaeological excavations have also revealed fortification structures and watchtowers in farmhouse architecture. One case is that of Farm 151, excavated by a joint Ukrainian–American team in the 1990s (Carter et al. 1997).

The archaeological record also suggests that eventually, after the Greeks became established in their farming colonies, the surrounding Scythian

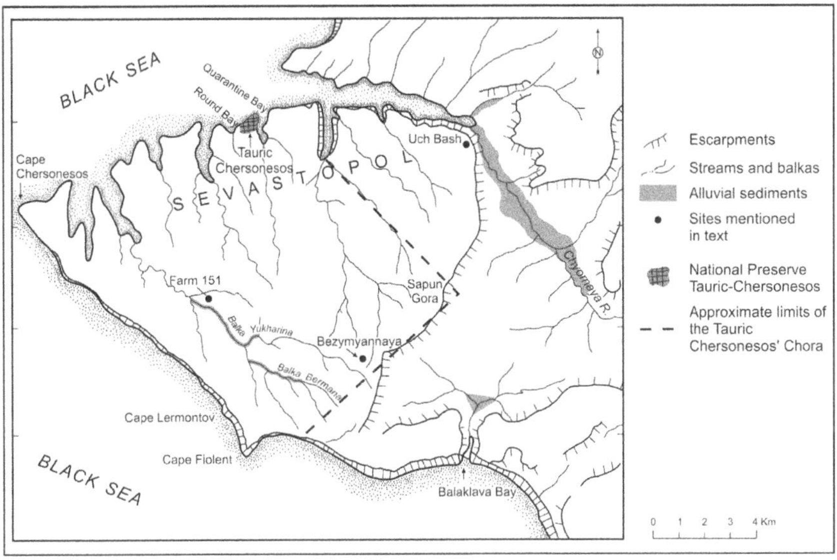

Figure 7.2: The chora of Tauric Chersonesos in the Heraklean Peninsula. Source: Cordova and Lehman (2003).

population became agriculturalists and began trading grain with the Greek colonists (Carter et al. 1997). Thus, much of the faming population in the rural territories around Theodosia (Figure 7.1) was Scythian, or, more properly, Hellenized Scythian (Gavrilov 2006). The settlement of Neapolis, the Crimean capital of the Scythians, in the suburbs of Simferopol, is an example of Hellenistic influences on a former nomadic group.

Ancient Greek viticulture

The initiation of grapevine cultivation in Crimea involved a process of adaptation to the local environment. Paleobotancial studies (Janushevich and Nikolaenko 1979; Yanuchevich, Nikolaenko and Kuzminova 1985) show that the Greek colonists domesticated the Crimean wild varieties of grape, rather than introduced Mediterranean grape varieties directly. That their decision to do so was the right one is supported by the length/breadth index measured on grape pips recovered from excavation of farms of different ages. The index varies through time: from the local wild grape variety of *Vitis sylvestris* to its domesticated form known as *Vitis vinifera*. Zoya Janushevich and Galina Nikolaenko (1979) support the idea that the local wild grape variety was better adapted to the lower temperatures of Crimea than the varieties from the Mediterranean. However, it is not clear whether this was achieved through grafting between imported and local varieties.

Cereals such as wheat, barley, and rye were also cultivated. However, Janushevich and Nikolaenko (1979) suggest that these crops were grown mainly in the northern districts of Chersonesos, along the west coast, where presumably they are better adapted to chernozem soils. Nonetheless, the presence of Ceralia-type pollen in the Chyornaya Valley pollen core suggests their cultivation in deep, hard soils of the alluvial plain and adjacent fluvial terraces (Cordova and Lehman 2003, 2005).

Other traditional horticultural crops of the Mediterranean were not always easy to adapt and manage in Crimea or in other areas in the northern Black Sea region. In particular, the cold-sensitive olive (*Olea europaea*) was never successfully introduced in Crimea in antiquity, and no evidence of olive cultivation appears in the pollen record for the area around Tauric Chersonesos (Cordova and Lehman 2003).

Other traditional Mediterranean crops, however, were easily adapted. Pulses were popular crops in the Chersonesite farms, as paleobotanical evidence indicates (Janushevich and Nikolaenko 1979). Other introduced cultivars of Mediterranean origin include figs, lavender and other essential oil plants, pomegranates and walnuts. Of all the introduced Mediterranean crops, the grape became the most important because of its adaptability to colder climates. This was achieved by grafting local wild varieties, and by a series of other techniques practised to protect the crop in winter. As Peter S. Pallas (1812) showed, the problem of adapting grapes to cold winters was solved by using a technique of planting the vines in holes, which were filled in with soil during the winter to protect them—a technique that lasted until the late eighteenth century. In fact, to this day, most of the grapevines are covered during the winter in most of the northern and central parts of the peninsula (Fursa 1977).

Climate, environment and ancient farming

Pollen diagrams from the northern coast of the Black Sea show a climatic amelioration in the middle and late part of the first half of the First Millennium BC—the time of rapid Greek colonization of the Black Sea. For the lower Dnieper region, this amelioration is reflected in the increase in tree pollen in the Kardashinski Swamp (Kremenetski 1995). A similar trend is evident in the pollen records of southeastern Ukraine (Gerasimenko 1997), an area close to the northern coast of the Azov Sea. This climatic amelioration is also apparent in the composite pollen record of southwestern Crimea (Figure 6.1) and the records of Lake Saki (Figure 6.4, top). During the same period a slight increase in tree pollen is evident in the pollen record of the Yaltinskaya Yaila (Figure 6.2). Furthermore, remains of oak and other trees were recovered from the Greek settlement at the Chaika site near ancient Kerkinitis, suggesting nearby stands of trees in the Tarkhankut Peninsula (Lyevkovskaya 1970). The finding is often taken as indicating improved conditions, and even as a source of timber for local construction and trade (Hannestad 2007). The area is dominated by steppe today, and receives less than the amount of precipitation required for tree growth. Nonetheless, the amount of rainfall is only barely below that of the Heraklean Peninsula, where, in its upper parts and valleys, stands of pubescent oak survive. Therefore, it is a possibility that oak trees thrived in the moister parts of the Tarkhankut Peninsula (possibly in valleys) during the relatively wet period that accompanied the Greek agricultural expansion of the fifth to fourth centuries. Another point that should be made is that if rainfed farming was possible, then most likely there was enough rainfall to maintain trees in some areas where the soils would hold moisture and where the trees could be protected from cold winds.

The sequence of farm establishment and expansion follows a complex combination of opportunities—some climatic, some ecological and some sociopolitical. In the context of a relatively favorable climate, Greek colonists first established their farms immediately around the urban areas. As relations with local tribes improved by trade or by force, Greek farming territories began to expand into the hinterland, which for most Crimean towns meant expanding into the steppe (Šceglov 2002; Shilik 1997).

The expansion of Greek territories occurred from the two main cities on the extremes of the peninsula: Panticapaion and Chersonesos (Figure 7.1). Then, acting as satellites of these main cities, other settlements and their own chorai sprung up (Saprykin 1994; Gavrilov 2006; Kutajsov 2006; Nikolaenko 2006). In the east, these new settlements included Nimphaios and Theodosia, and in the west, Kerkinitis (modern Yevpatoria) and Kalos Limen. Ultimately, the combined agricultural territories embraced the coastal areas of the plains region of the peninsula (Figure 7.1). Based on pollen diagrams, it seems that a relatively wet period existed during the middle of the First Millennium BC (Figure 6.1) when this colonization boom occurred.

The environmental and social history of the ancient town of Tauric Chersonesos—one of the most successful Greek colonies in the Black Sea—epitomizes the history of colonization in Crimea and the whole northern Black Sea region. The ruins of the ancient city of Tauric Chersonesos are located on the northern coast of the Heraklean Peninsula on a small peninsula between Round and Quarantine bays (namely, *Kruglaya Bukhta* and *Karantinaya Bukhta*) (Figure 7.2). The city of Sevastopol has engulfed the ruins, which today are part of the Tauric Chersonesos National Preserve and one of city's main attractions.

The foundation of Tauric Chersonesos has its roots in political events that occurred in the mother city of Heraklea Pontika, a Megarian city-state located in modern Eregli, in northwestern Turkey (Figure 1.4). A schism between its citizens forced one of the factions to seek a democratic form of government outside the city, forcing many of them to migrate northwards and found Chersonesos (Sorochan, Zubar and Marchenko 2000). The year of the colony's foundation is set at around 422/421 BC (Saprykin 1994). According to the ancient geographer, Strabo, the first settlement—referred to as "Old Chersonesos"—was located on the westernmost tip of the Heraklean Peninsula (Strabo 4, 2). Despite numerous early settlements in the peninsula, the location of this Old Chersonesos has been identified in the Chersones Peninsula, also known as Mys Mayachny (Lighthouse Point) (Figure 7.2) (Saprykin 1994; 2006). Later—perhaps in the fifth or fourth century—the city moved to its present location in the peninsula, next to Quarantine Bay. The first chora was limited to the Chersones Peninsula (Saprykin 1994; Nikolaenko 2006), but after the city was moved to its final location, the chora expanded to occupy most of the Heraklean Peninsula (Figure 7.2).

The Heraklean Peninsula consists of a limestone plateau tilted to the north and bordered by cliffs to the east and west. The plateau is dissected by a series of valleys known locally as *balki* (ravines), all of which empty into the northern shore of the peninsula (Figure 7.2). One of the main environmental factors that the Greek farmers faced in the Heraklean Peninsula was the thin calcareous soils (rendzina soils), which are the result of a widespread period of erosion that occurred during the Middle Holocene (Figure 6.1). But thin calcareous soils seem not to have deterred Greek farmers, such soils being the norm in most of the mountainous and calcareous lands of their Mediterranean homeland. Classical writers who wrote about Mediterranean practices, such as Lucius Junius Moderatus Columella, Marcus Terentius Varro, Marcus Porcius Cato and Pliny the Elder, explain in detail the adaptation of grapevine, olive and cereals, and other crops, to stony calcareous soils, as well as other typical facets of the Mediterranean landscape (Bradley 1725; Robinson 2006; Isager and Skydsgaard 1992; Hughes 1994). Locally, however, evidence of stone rows, retaining walls and barrages to manage soil and water are evidence of the number of techniques used to protect soil (Moysyeyev 1926; Strzhyelyetskiy 1961; Kumysh et al. 1999).

The cultivation of cereals was not widespread in the Heraklean Peninsula during the centuries of Greek farming. Although Janushevich and Nikolaenko (1979) recovered carbonized grains of wheat (*Triticum aestivo-compactum*) and rye (*Hordeum vulgare*) from excavated farms in the chora of Chersonesos, their findings suggest that cereal was not abundant in the chora's farms (Janushevich and Nikolaenko 1979). Frequencies of up to 2 percent of Cerealia-type pollen in zone 3 in the NG-2 core suggest the presence of cereal cultivation in the floodplain of the Chyornaya River (Cordova and Lehman

2003, 2005), but the numbers do not support major cultivation. Zoya Janushevich and Galina Nikolaenko (1979) found that cereal remains were more abundant in the farms north of the Heraklean Peninsula along the west coast, where soils were better suited for these crops than the soils of the western coast of the Heraklean Peninsula (Yanuchevitch, Nikolaenko and Kuzminova 1985; Cordova and Lehman 2003, 2006). Archaeobotanical remains from the farms in the Heraklean Peninsula consist mainly of grapes and pulses, which are better adapted to the local thin soils and warmer conditions of the south coast (Yanuchevitch, Nikolaenko and Kuzminova 1985).

The sequence of changes in the landscape of the chora of Tauric Chersonesos can be reconstructed using pollen obtained from the pre-Greek soil in the fortified farm at Bezymyannaya and the sediments of Balka Yukharina and Balka Bermana (Figure 7.2) (Cordova and Lehman 2003). The pollen sequences in the balka sediment show the presence of an open woodland in the period immediately before Greek settlement, with oak and pistachio, and some mesic trees growing most likely along the balki. The presence of willow and elm pollen suggests more moisture in the balki and perhaps some stream activity from water in springs. Small amounts of hazel and alder suggest mesic conditions provided mainly by springs and deep soils in the balki. Pollen from the sediments and soils coeval with the Greek colonization shows that tree pollen was reduced, and grasses and weeds expanded (Cordova and Lehman 2003). Thus, the establishment of the Greek farms transformed the open woodland vegetation and practically eliminated mesic microhabitats of the balki. Soils on the limestone plateau were deeply transformed as most of the land was turned into vineyards. Even today, rows of stones are seen in many localities near Balka Yukharina (Figure 7.2). Additionally, deep soils of the balki were transformed and dams and barrages were built to retain water and sediment that were subsequently plowed and cultivated (Moysyeyev 1926; Strzhyelyetskiy 1961).

The example of landscape change in the Heraklean Peninsula shows not only the modification of a landscape by local husbandry but also an abrupt change caused by adapting Mediterranean-type agriculture to an area that most likely had been used as pasture by the local natives. Thus, large tracts of the land were turned from pasture, and in some places woods, to vineyards and fields for horticulture. This change came with a number of stone structures, such as rows, fences, barrages at the bottom of balkas, roads, aqueducts and farmhouses.

The decline of Greek farming

By the end of the First Millennium BC, most independent cities of the northern Black Sea region had undergone decline. In some cases their populations and agricultural territories shrank; in other cases the cities were totally abandoned. Those settlements that remained lost their independence. This decline was accompanied by a series of wars with nomadic steppe peoples, sometimes inter-city wars, and in some cases the influence of Rome, which at that time began to expand its empire in the Black Sea Region (Sorochan, Zubar and Marchenko 2000). In Crimea, Greek settlement decline is nowhere else more obvious than in the shrinking of the farming territories of Tauric Chersonesos and the Bosporan Kingdom by the end of the second century BC (Maslennikov 1997; Sorochan, Zubar and Marchenko 2000; Šceglov 2002).

The idea of climatic deterioration as one of the factors responsible for the decline, and for the political and economic instability, is a recent one (Shilik 1997; Cordova et al. 2011). The palynological records of southwestern Crimea show tree pollen frequencies decreasing throughout the second half of the First Millennium BC, but this phenomenon was thought to come from the removal of trees as a result of the initial Greek farms in the Herkaklean Peninsula (Cordova and Lehman 2003, 2005). Nonetheless, the recent pollen data from the Lake Saki sediments provide evidence of a profound dry period between 2,200 years BP and 1,550 years BP (Gerasimenko 2011). This would have placed the beginning of the dry period around the third to second century, which is when most of the data suggest that the ancient Greek economic decline in Crimea occurred (Cordova 2011).

Stratigraphic evidence from alluvial deposits and soils in Tarkhanskaya Balka, on the northeast corner of the Kerch Peninsula, shows that the stream incised a black soil, indicating a transition from a stable wet floodplain to a dry flash-flood regime, an event that has been correlated with the decline postulated by archaeological and historical evidence (Cordova et al. 2011). Nonetheless, the case of the Tarkhanskaya Balka is a single isolated case that may or may not be associated with climatic deterioration. The question of whether climate alone is responsible for prompting political instability and war is still not answered, given the sparse record. But if climatic deterioration was responsible, it was a short-lived one.

This decline did not mean the end of the Mediterranean rural tradition in Crimea. The Greek legacy of Mediterranean land-use survived not only in the form of features in the landscape (remains of farms, terraces, walls, dams, stone rows) but also in the continuity of some of the practices that remain in place to this day. The introduction and adaptation of Greek farming may have aided other Mediterranean settlers in southern Crimea—Romans and Byzantines, and later in the Middle Ages, the Genoese, and Turks. Furthermore, settlers from the north and east, such as the Goths, Alans, Khazars and Tatars, adopted the rural systems in place. Likewise, Russians settling in the south of the peninsula also adopted established rural systems and, in fact, promoted many of the Mediterranean crops. In one way or another, the Mediterranean rural legacy, heralded by the Greek colonists, survives to this day in the peninsula

Romans, Byzantines and the Migration Period

The strengthening of the Scythians and the decline of the Greek colonies in the last two centuries BC happened at the time when Roman power began to be felt in the Black Sea region. In the middle of the first century BC, the half-Greek half-Persian Mithidrates IV Eupator, ruler of the Pontian Kingdom (or Bosporian Kingdom) and conqueror of the whole Crimea and surrounding areas, was defeated by General Pompey, thus laying the way for the Roman domination of Crimea (King 2004). The Romans established a series of fortifications along the coast between Eupatoria (Yevpatoria) and Kerch. During this time, the city of Chersonesos remained independent, in what is referred to as a Roman protectorate (Andryeyev 1997). Under these conditions, the city prospered, despite the fact that the areas of dominance in the western part of the peninsula were lost. The areas formerly controlled by the Pontian Kingdom in the eastern part of the peninsula, on the other hand, never regained prosperity, as they remained open to attacks from nomadic tribes and almost continuous wars.

In the third century AD, as the Roman Empire began to decline in the west, Roman military power in the northern Black Sea region also began to wane. Then two major events ensued: one was successive invasions by "barbarian" groups from the Empire's periphery; the other was the rise of the eastern successor of the Roman Empire, Byzantium, whose rulers were determined to retain territories in the Black Sea region formerly held by Rome.

The successive invasions by the numerous tribes gave way to the period known in Crimean history and archaeology as *Byelikoye Pyeryesyeleniya Narodov* (literally, the Great Migration of Peoples), which in English is known as the Migration Period. In the whole of Europe this period spans approximately between 300 AD and 800 AD. But in Crimea, the prominent Crimean Anatoliy L. Yakobson (1970) places this period between the third and tenth centuries.

The groups that invaded Crimea during the Migration Period were of varying origins: Germanic (Goths), Iranian-speaking (Alans) and Turkic-speaking (Huns first, and then Khazars, Pechenegs and Polovtsy). Although, formally, the Migration Period encompasses the Early Middle Ages, invasions and settlement of foreign peoples in Crimea continued throughout the High and Late Middle Ages (roughly eleventh to fourteenth centuries). This second wave of immigrants included the Venetians and Genoese, and the Tatar-Mongols.

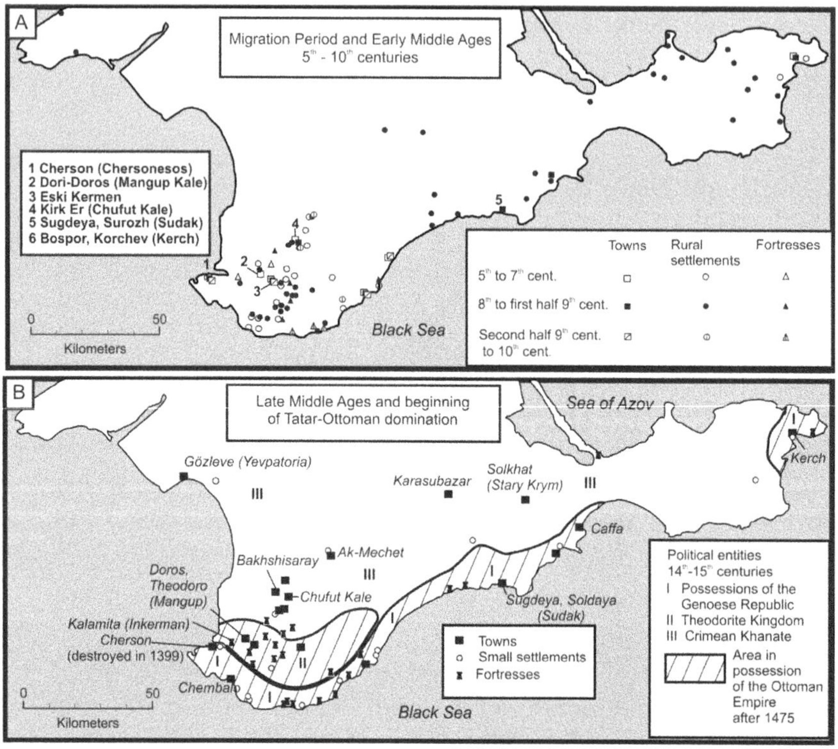

Figure 7.3: (A) Major settlements during the Migration Period and Early Middle Ages; (B) Main cities and polities during the Late Middle Ages and the beginning of Ottoman domination. Sources: Yakobson (1970) and Gyertsyen (2003) with modifications and additions by the author.

The immigrants of the Early Middle Ages occupied mainly the piedmont and mountains, as well as some areas of the eastern Kerch Peninsula (Figure 7.3A). The arriving peoples maintained for a while their cultural identity but eventually they merged with the existing Tauro-Scythian and Greek populations. In other cases the immigrant groups merged with one another, as was the case of the "Alano-Gothic," a term often mentioned in the sources (Yakobson 1970; Dyulichyev 1997). As most of the mountains and piedmont became settled by the newcomers, Byzantium struggled to retain control over the southwestern end of the peninsula, the south coast, and the eastern part of the Kerch Peninsula (Sorochan, Zubar and Marchenko 2000). Chersonesos became the seat and cultural center of Byzantine power in Crimea. In the east, Bospor (present Kerch) and other towns that had been destroyed by the Huns were revived under Byzantine protection, a process that also meant the construction of more fortifications (Dyulichyev 1997).

Two aspects should be emphasized with regard to landscape change in the Crimean Peninsula in the Migration Period. One is that a highly dense population area appeared in the piedmont and mountains, particularly in the western half (Figure 7.3A). The other is the perpetuation and strengthening of a cultural ecological duality characterized by a settled and agricultural south and a nomadic/semi-nomadic north.

The best documented of the peoples who settled in Crimea in the first phase of the Migration Period are the Goths, an eastern Germanic tribe whose territory extended from the Baltic to the western Ukrainian steppes (Vasiliev 1936). Their arrival in Crimea occurred in the third century AD, but their connection with the mainland Goths was severed by the Hun invasions in the fourth century AD, which is why they formed their own political entity centered in the mountainous southwestern part of the peninsula (Vasiliev 1936). The Goths adopted the Christian Orthodox faith and retained their ethnic and political identity until the fourteenth century, when they formed the Principality of Theodoro, whose capital stood on the Mangup Plateau. The principality eventually disappeared, as it was incorporated into the Crimean Khanate in the fifteenth century (Vasiliev 1936; Dyulichyev 1997).

The Khazars were another prominent group that arrived in Crimea towards the end of the Migration Period. Their center of domination was in the northern Caspian Sea, where they formed a confederation of Turkic-speaking tribes (Gyertsyen 2010). Although they maintained some sort of military and economic presence in Crimea, their final settlement in the peninsula occurred towards the end of the eighth century. They played an important role in the local political power balance, settlement, culture and socio-economic life of the peninsula. The fact that the Khazars dominated trade with the east and had for the most part good relations with the Byzantine Empire led to social and economic stability during the second half of the eighth century and through the ninth century (Andryeyev 1997; Gyertsyen 2010). This stability may have also accounted for the increase in population in the piedmont and mountains (Andryeyev 1997), although many aspects related to climate amelioration also played an important role in the population increase, as discussed in Chapter 6.

The Khazars converted to Judaism, probably through contact with Jewish populations who had settled in the marginal areas of the Byzantine Empire (Dyulichyev 1997). As a result, a unique Jewish culture flourished in Crimea, giving birth to Karaites and the Krymchaks, two sects whose descendants have survived until recent times in very small numbers, residing mainly outside Crimea (Dyulichyev 1997). The cultural landscape of this local Jewish culture is seen in the ruins of medieval settlements and cemeteries, particularly in and around Chufut Kale in the lower mountainous area of the Kacha River basin.

Some settlements founded by the migrating groups grew into towns, of which the most prominent were Mangup Kale, Eski Kerman and Chufut Kale (see locations in Figure 7.3A). These towns were surrounded by a number of smaller villages and farms now lost under a dense cover of forest. In these areas the use of natural resources, particularly soil and water for agriculture, became more intense than ever before (Schreg 2010, 2011). It is not clear why these mountainous areas were preferred, but Anatoliy L. Yakobson (1970) believes that the Hun invasions pushed people from the piedmont into the more protected areas of the mountains. But it also seems that there were other factors, related to resource availability (Schreg 2010) and, perhaps, suitable climatic conditions (see Chapter 6). The process of settlement in this region meant the removal of the forest, and the intensification of farming on the slopes had a strong impact on soil erosion, whose intensification goes in tandem with population increase and decline in this region (Schreg 2011). Recent studies involving survey, stratigraphy and geoarchaeological research have revealed an agricultural development system of soil terracing preserved under a dense forest cover (Schreg 2011).

Population development in the eastern part of the piedmont and mountains during the Early Middle Ages did not involve large towns and a high rural population density as in the western part of the peninsula (Figure 7.3A). It was not until the thirteenth century that the eastern part of the mountains and piedmont region saw an increase in population and the rise of cities, particularly with the presence of the Genoese in Caffa and Sudak, and with the Tatar-Mongols in Stary Krym and Karasubazar (present-day Belogorsk) (Figure 7.3B).

Towards the end of the Migration Period, other groups appeared in the peninsula. The Kievan Rus established a trading post on the Taman Peninsula and eventually dominated Bospor in the Kerch Peninsula. The Rus established economic ties with the Byzantines via Chersonesos. This relation resulted in the Christianization of the Kievan Rus and the strengthening of relations between the two states, which allowed trade between Byzantium and the Baltic regions.

In the tenth century, the Pechenegs, a Turkic-speaking nomadic group that had been raiding the Kievan Rus territories since the eighth century, seized the Tmutarakan Kingdom from the Kievan Rus in the tenth century (Dyulichyev 1997). This disrupted many of the relations between the Rus and the Byzantines, and compelled the latter to reinforce their fortresses along the coast. But after the Pechenegs became established, they were challenged by another nomadic tribe, the Polovtsy (known also as Kipchak), who eventually took over the eastern part of Crimea, establishing themselves in Sudak, where they extracted protection tribute from neighboring polities including

Chersonesos, the Theodorite state, and later the newly established Genoese colonies (Figure 7.3B).

The High and Late Middle Ages

Venetian and Genoese merchants, whose rivalry had involved competition for markets in the Mediterranean, and later the Black Sea, since the eleventh century, began to compete for posts in the Crimean Peninsula, a territory located at a strategic point in the trade with Asia (Dyulichyev 1997). Eventually, the Genoese succeeded in establishing themselves in several locations on the south coast, principally at Caffa (Kaffa) in modern-day Feodosia, and in extending their forts and centers of trade along the south coast from Caffa in the east to Chembalo (present-day Balaklava) in the west (Dyulichyev 1997). Ruins of these fortifications can still be seen, particularly the well-preserved Sudak Fortress. In the subsequent centuries the Genoese would play an important role in regional trade, and also in linking Crimea with other regions of the world.

Polovtsy dominance in eastern Crimea was disrupted in the eleventh century by the invasion of another Turkic-speaking group, the Mongols, who soon displaced them and eventually came to dominate the entire peninsula. The expansion of the Golden Horde in Russia and eastern Europe began with the catastrophic Mongol raids of the early 1200s. The first Mongol raid of Crimea occurred in 1223, during which most towns were sacked and burned, including Chersonesos (Dyulichyev 1997). Sudak, the seat of Polovtsy domination, was leveled to the ground (Andryeyev 1997). But after this initial attack, the Mongols did not stay. It was not until a raid in 1239 that they seized parts of the peninsula, particularly what used to be Polovtsy territory. They established their seat of power in Stary Krym in the valley of the Churuk-Su River in the eastern piedmont (Figure 7.3B). Crimea became part of the Golden Horde, an empire that extended over a vast area of Eurasia, from China to Kiev and from northern Russia to Iran. In the historical sources, the group that became established in Crimea was known as the Tatar-Mongols, sometimes referred to simply as the Tatars.

Relations between the Tatar-Mongols and the Genoese were for the most part uneasy and violent, but eventually a series of accords allowed the Genoese to continue their trading enterprise at the port of Caffa, which became one of the main points on the long trade route between Europe and East Asia, namely the Silk Road. Although this trade route allowed contact between the two ends of the Eurasian continent, including the exchange of ideas, invention and goods, it also brought the beginning of a catastrophe: the Black Death.

After its outbreak in China, the bubonic plague travelled west with the merchant caravans, reaching Crimea in 1346. Unaware of its true origin, the Tatar-Mongols believed the plague had been spread by the Genoese, which is one reason they launched an attack on Caffa. The sources seem to vary, but there are accounts that the Tatar-Mongols catapulted infected dead bodies into the besieged city (Wheelis 2002). Genoese refugees escaping the siege by sea spread the plague rapidly through the ports along the Bosporus and the Mediterranean (Wheelis 2002). The plague first reached Sicily and Genoa in 1348, and then spread rapidly across Europe, where in the next few years it devastated a large portion of the western European population. From Constantinople the plague also traveled to other areas such as Egypt and the Near East. By land, the plague spread through Ukraine and Russia. The number of deaths in Crimea is unknown, although it was probably as devastating as the Justinian Plague a few centuries earlier.

The post-medieval peninsula: Tatar and Ottoman rule

The Crimean Tatar polity was part of the Golden Horde, which comprised a complex tapestry of tribes, divisions and factions. In the early 1400s the central power of the Golden Horde began to dwindle as rivalries among factions began to break away, leading to the disintegration of the empire into various smaller political entities. In this process, an independent Crimean Tatar state was created in 1443 under the leadership of Haci I Giray, who became the first Crimean khan. However, the independent Crimean khanate did not last too long. The rise of the Ottomans after the collapse of the Byzantine Empire meant an increase of the Sultan's influence on Crimea's affairs. After internal strife among the Tatar rulers, the Ottomans imprisoned Mengli I Giray, the son of Haci I, and made the Crimean khanate a protectorate—a status that lasted until the Ottoman–Russian War of 1768–74.

The Crimean khanate and the Ottomans joined efforts to oust the Genoese from the region. The final blow to the Genoese posts in Crimea occurred when the Ottomans invaded the south coast in 1478. The expulsion of the Genoese from Crimea, and the takeover of most lands around the Black Sea by the Ottoman Empire, severed the trade routes between Asia and Europe. It is after this event that Europeans began their search for alternative sea routes to the Orient—which led to the age of discoveries overseas.

The Ottomans took direct control of the lands formerly controlled by the Genoese on the south coast, as well as the territories previously held by the Theodorite Kingdom and Chersonesos (destroyed in 1399), and the Kerch Peninsula (Andryeyev 1997). The Crimean khanate kept the rest of the peninsula and vast areas of steppe between the Dnieper and the Don, and east of the Azov Sea. Beyond these lands a great expanse of no-man's land divided the Crimean khanate itself from the lands under the domination of Muscovy, a state that later became the Russian Empire.

The areas dominated by the Ottomans in Crimea became the *sancak* of Kefe, which also included areas across the Strait of Kerch, namely the Taman Peninsula, and areas around the Azov Sea. The economy of Ottoman Crimea has for some time been a matter of debate among historians, particularly in relation to the issue of whether the Ottomans used the Genoese trade connections. Some historians believe that trade in the northern part of the Black Sea declined after the Genoese left, but a search through Ottoman documents revealed that, for at least two centuries, trade centered at Caffa continued (Veinstein 1986). However, modifications existed in terms of commodities. For example, the port became an important hub for the slave trade, linked to the raids Tatars carried out in the neighboring lands of Lithuania, Poland and Muscovy (Veinstein 1986). In terms of local products for export, a number of items are listed, most of which are agricultural products, honey and fish. Interestingly, however, despite prohibition of alcohol, the wine trade in the areas around Black Sea continued. Through a study of Ottoman tax documents, Oleksander Halenko (2004) concluded that the profitable wine taxation revenue on Christian producers accommodated wine production to Muslim principles by promoting conversion to Islam. In the records, the Crimean districts with the largest tax revenue in the sixteenth century were Sudak, Caffa and Inkerman.

Ottoman trade in the northern part of the Black Sea began to decline by the end of the seventeenth century. Although the causes of this decline are debatable, one reason that has been put forward is the strengthening of the expanding Russian state to the south under the rule of Peter the Great (1682–1725). But even before the expansion of Russia beyond the northern borders of the Ottoman Empire, Cossack naval raids on Ottoman ports and shipping had already begun to disrupt trade, as noted by Italian missionary Emidio Portelli d'Ascoli in his *Description of the Black Sea and Tartary* in the first half of the seventeenth century (Ostapchuk 2001).

The imprint of Tatar and Ottoman cultures in the landscape

Closely related linguistically, and later by the Islamic religion, the Ottoman Turks and the Crimean Tatars developed a close relation that seems to have lasted through most of the period of Ottoman influence in Crimea. It seems, nonetheless, that Tatar culture survived, absorbing Mediterranean and Central Asian elements. Thus, Tatar culture developed as an identity which has been carried over to this day.

During the four centuries of Tatar domination, a Crimean Tatar culture developed in the regions where they established their centers of power. The first center was in the eastern piedmont, in the town of Stary Krym. Later, their center of power moved to the western piedmont, first to the town of Chufut Kale, and finally to Bakhshisaray (Figure 7.3B).

The name Stary Krym is the Russian translation of the Tatar *Eski Qirim*, meaning Old Crimea. The medieval town was known as Qirim, which is believed to be the origin of the name Krym, or Crimea. The town was in the fertile valley of the Churuk-Su River, at a key location with easy access to the steppe and the sea. The area around this town became settled predominantly by Tatars, whose former life had been nomadic, but turned into a more established way of life practising horticulture and pastoralism (Dyulichyev 1997). When the capital moved to the western piedmont (Chufut Kale and later Bakhshisaray), another area of Tatar population and culture flourished there as well. Eventually the entire piedmont, some valleys in the mountains, and the south coast, were settled by Tatars, whose main livelihood depended on horticulture, apiculture and grain farming, among other rural activities. The only sector of the Tatar population that remained nomadic were the Nogay Tatars, who practised pastoralism in the steppes of northern Crimea and the vast steppe areas of the khanate in the mainland.

Towards the end of the Ottoman–Tatar period, most of Crimea's population had become a melting pot where local Tauri, Greeks, Goths, Alans, Tatar-Mongols and Italians blended into a single population nationality recognized as Crimean Tatar, which was held together by the Crimean Tatar language and the Islamic religion. But within this unity, some cultural differences remained. Peter S. Pallas (1812) noted strong differences in customs, dress, way of life and other cultural traits. Pallas also became interested in the differences in dialectal forms of the Crimean Tatar language, which he was able to differentiate by regions, distinguishing in particular the dialect of the south coast from the rest of the peninsula.

As with the rest of the Ottoman lands, Crimea was home to a varied number of non-Muslim minorities who lived in relative harmony with the Muslim majority. The Christians in Ottoman-Tatar Crimea were composed mainly of Greeks, Armenians and Georgians. The Jewish population comprised predominantly the Karaite and Krymchaks, whose origins date back to the Early Middle Ages. Most of the Christian and Jewish populations in Crimea were townspeople engaged mainly in trade and small industry (Andryeyev 1997).

During Ottoman-Tatar times, small villages were abundant, particularly along valleys in the mountains, the piedmont, and along the south coast, where there existed a mixed rural economy based on meat-dairy production, horticulture and, in some areas, grain cultivation (Shumskiy 2009). The plains were devoted mainly to dairy and meat production linked to nomadic pastoralism (Dyulichyev 1997; Shumskiy 2009). The rural landscape was dominated by the feudal structure established in the early days of the khanate. The land was divided into districts dominated by the *bey*, which in turn assigned the land among themselves and the *mirza* (the local feudal lords). The villagers were serfs who owned no land and worked for their lords. Additionally, some lands were directly owned by the khan, including mainly forested areas in the river valleys, a number of territories in the piedmont, and the salt lakes in the north (Dyulichyev 1997). In the areas dominated by the Ottomans, the land was under the direct control of the Ottoman sultan, where rural production was mainly animal goods, fruits, honey, cereal, table grapes and dairy and meat products (Dyulichyev 1997).

By the end of the Ottoman-Tatar domination, in the late 1700s, Crimea had several cities. In the lands dominated by the khanate, the largest towns were Bakhshisaray and Ak-Mechet (present-day Simferopol). The only port under the control of the khanate was Gözleve (present-day Yevpatoria). The major city in the Ottoman lands of the peninsula was Caffa (Feodosia), which was the main port in Crimea. Kerch was another port, but less important than Feodosia. The rest of the population centers were small towns and villages. The most prominent imprint left by Tatar-Ottoman rule in the cities was the architectural monuments, in particular mosques and palaces, of which the most impressive is the Khan's Palace in Bakhshisaray, which still stands today. Another important legacy of the Tatar and Ottoman period is perhaps the large number of place names in the landscape—more than those from all the other linguistic influences (i.e., Greek, Russian and Ukrainian).

CHAPTER 8

CRIMEA IN LATER HISTORY AND MODERN TIMES

Annexation to the Russian Empire

Czar Peter the Great (1672–1725), also known as Peter I, initiated the long process of Russian expansion southward. His objective was to procure an ice-free port for Russia in the Black Sea, which meant a direct confrontation with the Ottoman Empire. After various wars with the Ottomans, however, Peter's gains reached only as far as the mouth of Don River at the northern end of the Azov Sea. His successors, particularly Catherine the Great (Catherine II), continued the fight over access to the Black Sea. During her reign, the forces of the Russian and Ottoman Empires were especially put to the test during the second half of the eighteenth century.

The war of 1768–74 tipped the balance of power in the northern Black Sea region in favor of Russia, with Crimea at the center of the confrontation. As Russian military gains grew at the expense of Ottoman losses, so feelings of Tatar identity within a sector of the Crimean nobility led them to seek independence from the Ottomans under the protection of Russia (Fisher 1978). Thus, a treaty with the Russians in early 1772 made the Crimean khanate an independent state under the protection of the Russian Empire. With Russian military presence in Crimea, a large sector of the pro-Ottoman nobility, particularly the *mirza*, fled to Istanbul with the aim of reorganizing a re-conquest of the khanate (Fisher 1978).

Ottoman attempts to re-conquer Crimea, together with political division within the pro-Russian khanate factions, prompted Russia to a full military occupation of the peninsula and a declaration of annexation to the Russian Empire on 8 April 1783. Empress Catherine II appointed Prince Grigoriy Aleksandrovich Potemkin governor of the newly acquired lands, which included not only Crimea but a vast territory in the northern Black Sea region. The Crimean Peninsula became the Taurian District (*Tavricheskaya*

Oblast'), which was part of the Novorossiysk Governorate (*Novorosiyskaya Guberniya*) until 1804, when it became the Taurian Governorate (*Tavricheskaya Guberniya*).

The land under Russian and Soviet rule

Along with the establishment of Russian military power in the Crimean Peninsula, a process of political, economic and cultural transformation began soon after official annexation. This transformation in turn meant the development of agriculture and industries, as well as the creation of new cities. Simferopol was built as a new city next to the Tatar town of Ak-Mechet. In 1784 the port of Sevastopol was founded in one of the natural harbors of the Heraklean Peninsula with the purpose of becoming the base of the Russian Navy in Crimea. The harbor was not only deep enough to handle large ships: it was also ice-free in winter.

The large number of Tatars who fled Crimea during the wars and annexation left a population vacuum in the cities and the countryside. In the cities, many of the abandoned houses were used temporarily as barracks by the occupying army, but still the number of abandoned houses, some of which were already derelict, amounted to several dozens in cities such as Yevpatoria, Ak-Mechet and Caffa (Druzhinina 1959). A similar situation occurred in the countryside, where villages were left partially or completely abandoned (Dyulichyev 1997). To fill this population vacuum, Prince Potemkin began a campaign to attract immigrants to the peninsula. As part of this campaign, Feodosia and Sevastopol were declared open cities for any immigrants (Andryeyev 1997). In the countryside, particularly in the areas dominated by the Ottomans, lands were also open for settlers. Most of the immigrants who settled in Crimea during this period were Russian and Ukrainian, but a large number of other nationalities settled as well, including Germans, Armenians, Georgians, Bulgarians; Greeks from areas dominated by the Ottomans also constituted a large number of immigrants (Fisher 1978; Dyulichyev 1997). The Germans, who had already been emigrating to Russia during Catherine's rule, established several colonies around Simferopol and Feodosia (Druzhinina 1959). In the end, the post-annexation immigration raised the peninsula's population from 60,000 in 1783 to 100,000 by the end of the century (Andryeyev 1997).

Those members of the Tatar nobility who stayed were incorporated into the Russian nobility and allowed to continue the semi-feudal ownership of their land (Druzhinina 1959). Lands deserted by former Tatar and Ottoman tenants were offered as land grants to the Russian nobility, some of whom moved to Crimea with their peasant serfs (Druzhinina 1959). The peasants who migrated to Crimea were predominantly Russian and Ukrainian and made up 51 percent (45,702 persons) of the total migrant population between 1783 and 1854 (Dyulichyev 1997).

Although the Russian population did not become the largest group until after the Crimean War (1853–4), the beginning of the nineteenth century saw Russian begin to dominate as the main language of commerce, administration and instruction, and overall the language of communication, to the point that the non-Russian population began to adopt it as their main language. Of all the ethnic groups in the peninsula, only the Tatars retained their language.

Karl Hablitz's report on the natural resources of the Crimean Peninsula was finalized and submitted to Prince Potemkin in 1785. Shortly after, the prince began to devise a plan for developing the countryside in the peninsula, basically by encouraging agriculture, horticulture and viticulture, for which he called on western European specialists (Shamko 1963). To enable development, the existing roads were improved and new roads were built, as is the case of the Simferopol–Yalta road (Dyulichyev 1997). Ports were re-designed and improved: Feodosia, Kerch and Yevpatoria were developed into commercial ports, whilst Sevastopol was developed into a port exclusively for the use of the navy (Andryeyev 1997).

The nineteenth century saw rapid economic development and population growth. Viticulture was developed (see section below), horticulture was expanded and, overall, agriculture was modernized. Tobacco, olives and other subtropical cultivars were introduced under the technical auspices of the Nikitskiy Botanical Gardens (established in 1812). Although agriculture increased and diversified, it was destined only to satisfy local markets (Andryeyev 1997; Shumskiy 2009). Small manufacturing enterprises were established in many areas, but never in the proportions or with the diversification achieved in other parts of European Russia (Shumskiy 2009). Industries focused mainly on construction materials and textiles, and in general employed a very small proportion of the population (Dyulichyev 1997; Shumskiy 2009). The lack of a railway system was a limitation on industry, agriculture and exports. Travel between the rest of the Empire and Crimea was usually done by sea via Rostov-na-Donu in the Azov Sea and Odessa. This lack of fast land transport proved a particularly detrimental factor in the supply of troops and materiel to the front during the Crimean War.

The most rapid development of agriculture, industry and commerce did not occur until after 1860. The main reasons for this new phase of development had to do with three significant changes in Crimea and the Russian Empire. First, land reforms in Russia cleared the way for modern capitalist enterprises, allowing the immigration of large numbers of freed peasants from the mainland

(Shumskiy 2009). Second, the treaty that concluded the Crimean War contained limitations on Russian naval activities in the Black Sea—one reason why demobilized navy personnel swelled the ranks of skilled labor in commercial activities and why Sevastopol became a commercial port also (Shumskiy 2009). The relevant clause of the treaty, however, was revoked in the 1880s, after which Sevastopol became once more a naval base, although it remained a commercial port also. The third reason for the new phase of development was the construction of the railway, begun in the 1860s, which stimulated agriculture, industry and commerce. The Zaparozhye–Simferopol line was inaugurated in 1874, the Simferopol–Sevastopol line in 1874 and the Dzhankoy–Kerch line in 1892 (Dyulichyev 1997). The main cities and ports of Crimea were then connected with the rest of the Russian Empire.

The fastest growth of the Crimean economy in imperial times took place between 1885 and 1913, in part due to the growth of capitalist monopolies (Shumskiy 2009). Manufacturing, viticulture and horticulture expanded, and the newly established tourism industry began to develop apace. The areas of major agricultural and industrial development were usually closer to the ports, particularly Feodosia, Sevastopol and Kerch, or in the piedmont region around the capital.

Little agricultural development, however, took place in the steppes. The main problem there was the lack of reliable precipitation and the difficulties in implementing irrigation given the water resources of the peninsula and the existing technology. Therefore, most of the steppe remained as pasture, and the areas near the Sivash and Perekop were used mainly for salt mining.

The twentieth century saw the fastest and most profound transformation of the landscape in Crimea, particularly during the Soviet period (1918–91). Not only did collectivization change many aspects of the rural and urban landscape, but the infrastructure also created opportunities for mass tourism and irrigated agriculture. Four phases may be distinguished during this period (Shumskiy 2009). In the first phase, roughly between 1918 and 1928, economic growth was very slow due to the many problems caused by the Civil War (mainly acts of sabotage), the implementation of collectivization, and numerous political problems. The second phase, between 1928 and the 1940s, was characterized by an increase in manufacturing linked to the mining industry, metallurgy and shipbuilding; this phase was interrupted abruptly by World War II. The third phase, encompassing most of the 1950s, was characterized by rebuilding and the reorganization of the economy. The fourth phase, from the early 1960s to the late 1980s, was characterized by rapid development in all fields: viticulture, horticulture and the construction industry, as well as tourism. This phase saw also the development of irrigated agriculture in the plains aided by the construction of the North Crimean Canal (Figure 8.1).

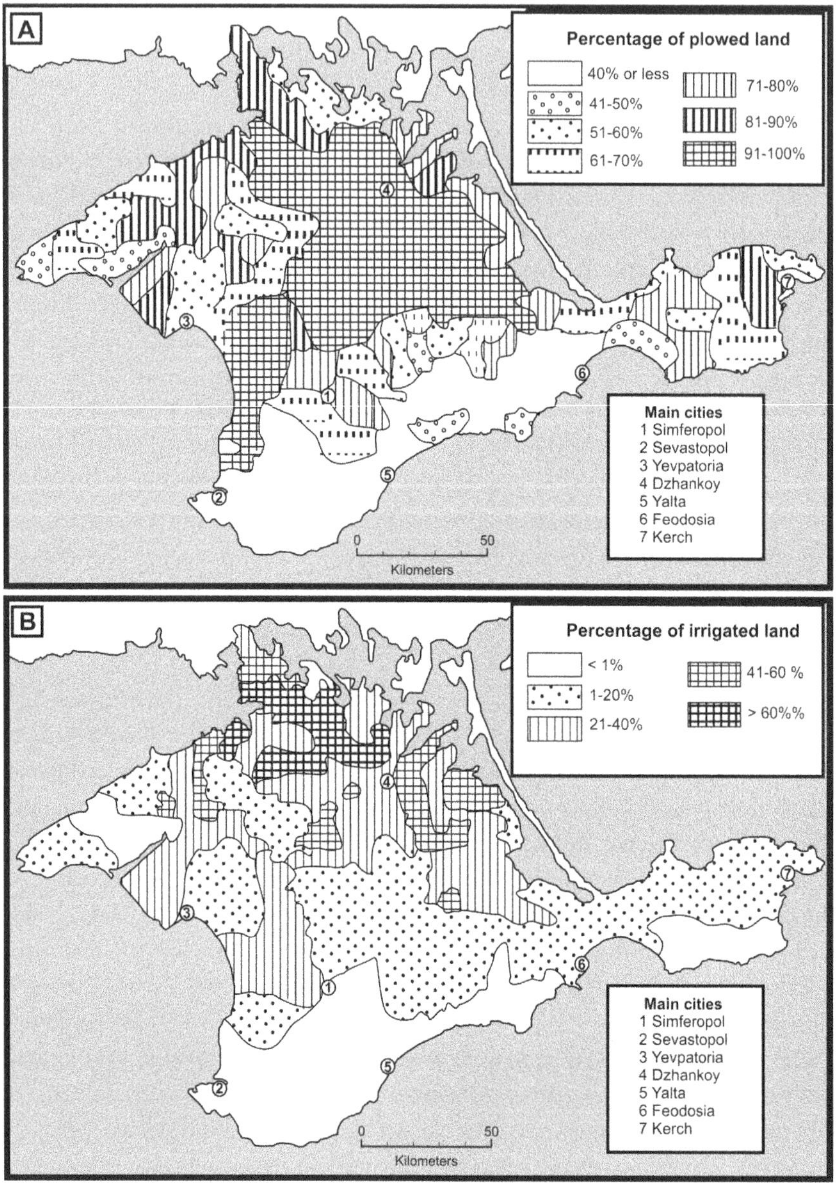

Figure 8.1: (A) Percentage of plowed land and (B) percentage of irrigated land in the Crimean plains. Source: Dragan (2004).

The mass deportation of the Tatar population in 1944 meant changes in population and land-use, which in turn transformed the cultural landscape, as discussed below. Another event of political importance during the Soviet period was the transfer of Crimea from the Russian Federation to the Soviet Socialist Republic of Ukraine in 1954. The reasons for the transfer are still debated but, among them, it is mentioned that Nikita Khrushchev was of the opinion that geographically and administratively Crimea was closer to Ukraine, and that the transfer was a reward for Ukraine's participation in the liberation of the country from the Nazi invaders and for its production of grain (Fisher 1978). However, officially the transfer was an action to commemorate the 300th anniversary of the Ukrainian union with Russia, so that Russia gave Ukraine the Crimean Peninsula as a sign of trust between the two Soviet partner states. At the time, the transfer created little concern among politicians and the public, since Ukraine was part of the Soviet Union (Fisher 1978). It was only after the collapse of the Soviet Union and the ensuing independence of Ukraine in 1991 that geographical and political issues appeared in relation to the transfer. These issues, which, among other political and economic problems, included the Russian naval bases in the peninsula and the discontent of the ethnic Russian population over the use of the Ukrainian language, led to radical opposition to Kiev and growing pro-Russian feelings. Under this climate of pro-Russian discontent, the Euro-Maidan events in Kiev in late 2013 and the ousting of President Viktor Yanukovich led first to the declaration of independence and a referendum to join the Russian Federation in March 2014.

Landscapes of war and peace

The Crimean Peninsula occupies a strategic position on the Black Sea. On the one hand it acts as a spearhead into the Black Sea; on the other, a fortress island guarding the northern Black Sea shores. Furthermore, the Kerch Strait is not only a strategic connection point between the Azov and Black Sea, but also an important place for shipping between the Mediterranean and the Don River. In other words, whoever dominates Crimea, dominates the northern half of the Black Sea and the Azov Sea. This location not only motivated imperial powers to establish military bases on the peninsula, but it was also at the root of many armed conflicts.

Since ancient times, fortresses have been built and rebuilt, and the militarization of the country has been such that remains of that period are found almost everywhere in the peninsula. The presence of ancient trenches and fortifications in the Perekop, the Arabatskaya Strelka, and the Kerch Peninsula are reminders of the defensive purposes of settlement in Crimea. The line of fortresses along the south coast, established by the Romans and Byzantines and later re-fortified by the Genoese, is another example of the military importance of the peninsula.

The succession of defensive posts at strategic points can be traced over several millennia. One example is the Bezymyannaya Hill in the Heraklean Peninsula, where given its high position, archaeologists have found in the same spot successive occupations linked to a fortified post from ancient Greek times to the time of World War II (Carter et al. 1997). A similar situation is seen at Kara Tobe, a fortress by Lake Saki facing the sea. A pillbox built during World War II stands on the ruins of an ancient fortress used by Greeks Scythians and Romans. Cases of such fortifications superimposed on earlier ones abound across Crimea.

Despite the long history of military conflict in Crimea, those that have left a greater impact on the landscape are the Crimean War and the Great Patriotic War (1941–5), as World War II is known in the territories of the former Soviet Union. Although to a lesser degree, World War I (1914–18), the Russian Civil War (1918–22) and the Cold War (1945–91) were also significant in their impact on the making of the modern landscape.

For the study of landscape change, the wars of the nineteenth and twentieth centuries have left a legacy of resources in the form of documents, particularly descriptions of the landscape and photographs. Among the documents, the most important are the maps, particularly those produced during the Crimean War, some of which show the extent of the forested areas, wetlands and other features existing at that time. The cartographic information of these features is complemented by later photographs.

Many descriptions rendered by reporters and writers are useful. As a young officer and correspondent, Leo Tolstoy remained in Sevastopol during the siege. His book, *Sebastopol Sketches*, describes many aspects of the landscape around Sevastopol, in particular a description of the Chyornaya Valley, which was a no-man's land between Russian and French positions (Tolstoy 1986). In the narrative and a poem, he describes the marshy nature of the floodplain, which is now drained and plowed.

Photographs taken by Luftwaffe reconnaissance planes during the invasion and siege (September 1941–May 1942) are a great resource, particularly for vegetation and landscape reconstruction in areas now occupied by urban sprawl and orchards. These photographs were very helpful, for example, in showing the morphology of the valleys of the Heraklean Peninsula for a geoarchaeological study (Cordova and Lehman 2005; Cordova et al. 2011), since most of these valleys have been modified by settlers, particularly by removing sediment for gardens around their dachas. Likewise, *Corona* satellite images taken by American spy satellites during the Cold War have been useful to archaeologists mapping the division lines of ancient Greek farms in areas now occupied by the sprawl of Sevastopol (Carter et al. 1997).

The damage to the vegetation during the armed conflicts of the nineteenth and twentieth centuries can be evaluated by looking at battlefield photographs. During the Crimean War, photographs taken by the British near Balaklava and other parts of the Heraklean Peninsula show completely destroyed woodlands in areas now forested with oaks. In part this destruction was caused by the cutting of the trees for firewood. Similarly, burned and destroyed forests are evident in World War II photographs. These photographs show that most of the trees in the woodlands of the Heraklean Peninsula, the Inkerman Valley and the Mackenzie Hills are not older than 70 years.

The Cold War saw not only the establishment and expansion of military bases but also the closing of many areas to foreigners and even to the average Soviet citizen. Many towns and areas that became closed zones or closed towns reopened only in the 1990s. The establishment of military bases represents a change in the landscape, both for bad and for good. On the one hand, some areas were used for training, which meant degradation of the landscape. On the other hand, the existence of bases meant protection of the land from other activities such as agricultural and urban development. In essence, through this type of protection, some of the existing bases or the areas around them have indirectly functioned as nature reserves. With the post-Soviet demilitarization of former bases, some of these territories are becoming *de facto* nature reserves (see the example of the Karan' plateau in Chapter 10).

Tourism and the Crimean landscape

The most important factor in the initial development of tourism in Crimea was the mild Sub-Mediterranean climate of the south coast, which was one of the warmest places of the Russian Empire. Other important factors were the scenery, hunting and fishing, and the curative properties of the mud from saline lakes in the plains.

Throughout the nineteenth century, the Russian nobility built palaces of retreat and mansions on the south coast, a trend initiated by count Mikhail S. Vorontsov in the 1820s. Other groups of the privileged classes, including a sector of the intelligentsia, began spending time in Crimea, generally for curative purposes. Based on the curative properties of the Crimean climate and the lake muds, sanatoria, or health resorts, began to appear at the end of the nineteenth century. The style of the hotels and resorts began to follow that of the Mediterranean coasts of France and Italy, one reason why the south coast of Crimea was often referred to as the *Trieste of the Black Sea* and the *Russian Riviera.* The center at the end of the nineteenth century was the Alupka-Yalta region. Thus, from being a tiny village of only eight houses in 1805 Yalta grew into one of the largest towns in Crimea at the beginning of the twentieth century (Dyulichyev 1997).

Mass tourism was not seen until after the Bolshevik Revolution, and more properly until the second half of the twentieth century. During the Soviet period, the tourism industry was developed both for party officials, some of whom made use of the dachas and palaces that had previously been constructed by the nobility and by the worker unions. Youth tourism also developed after the late 1960s. For example, Artyek, a children's camp, near Alushta, is to this day a popular destination for children and youth groups.

Tourism saw a sudden decline after the collapse of the Soviet Union—not surprising given that most of the visitors originated in the former Soviet state—but during the first decade of the twenty-first century this industry began to flourish once more. Today, not only has the number of national tourists increased, but so also has the number from other former Soviet republics, principally Russia. Visitors from other countries have also been increasing but only steadily.

The post-Soviet renaissance of the tourism industry has also seen many turns, as new resorts and areas of interest have become popular. Those cities formerly restricted because of their military installations, such as Sevastopol and Balaklava, have now become important tourist attractions. Despite the development of tourism in other parts of the peninsula, the south coast remains the center of most of this industry, accounting for the 48 percent of the total tourism infrastructure in Crimea (Yakovyenko 2009).

As with other economic activities, the rapid growth in tourism has taken a toll on the environment. Along the south coast, the expansion of tourism infrastructure, along with viticultural and horticultural fields, has left fragmented territories of Sub-Mediterranean vegetation, encroaching and threatening rare and endangered species (Biodiversity Support Program 1999). In addition to the growth of infrastructure, the lack of control of the ever increasing number of tourists visiting the forests and the yailas is threatening the highly vulnerable ecosystems and their endemics and endangered species.

The viticulture industry

Viticulture in the Crimean Peninsula did not die with the collapse of Greek farming. It gained importance again in the Early Middle Ages as a number of areas began to produce grapes, particularly in the piedmont and mountains (Rybintsev 1995; Shamko 1963). The spread of Christianity in Byzantine Crimea helped the expansion in cultivation, particularly through the establishment of monasteries. Furthermore, the Genoese established vineyards around their main colonies on the south coast, particularly in the Sudak area (Shamko 1963; Rybintsev 1995).

A decline in grape cultivation and wine production occurred during the Tatar and Ottoman domination, as the consumption of alcoholic beverages is prohibited by Islam. Vineyards were limited to the production of table grapes, and only monasteries and the Karaite Jews were allowed to produce wine

(Shamko 1963; Dyulichyev 1997). Based on western traveler accounts in the sixteenth, seventeenth and eighteenth centuries, the extent of vineyards in Crimea was very limited (Shamko 1963). The region of the south coast between Yalta and Sudak, which today is almost all covered by vineyards, was in the late 1700s devoted to grain agriculture (Shumskiy 2009).

In the late 1780s, Prince Potemkin brought in French advisors for the establishment of the wine industry, and soon new varieties were introduced and experiments were carried out with them (Shamko 1963). In 1801, an imperial decree established state sponsorship of viticulture, which consisted of state-owned vineyards that eventually would be transferred to individual owners (Rybinstev 1995). In 1804 a school of viticulture under state sponsorship opened in Sudak, and in 1828 the Magarach Viticultural Institute was established in Yalta (Rybintsev 1995). State efforts to develop the wine industry paid off, as annual production increased dramatically during the first half of the nineteenth century: from 102,000 buckets of wine in 1790, to 438,000 in 1837, to 600,000 (equivalent to 60 million liters) in 1867 (Rybintsev and Petkilyova 1992).

In the second half of the century more experiments were carried out to diversify the varieties of grapes and the varieties of wine, particularly champagne and cognac. The Russian oenologist Prince L'yev S. Golitsyn, after a number of experiments, began producing Crimean champagne in a factory that still exists today in Novy Svyet. Likewise, other specialties began to be developed in the various schools and companies. The extent of grape cultivation increased dramatically during the second half of the century, with vineyards occupying large areas in the south coast and piedmont. The total area rose from 5,095 ha in 1870 to 8,345 in 1880, and then to 11,030 ha in 1890 (Rybintsev and Petkilyova 1992).

Viticulture received strong support during the Soviet period, turning it into a major commercial activity. Pre-revolutionary wine enterprises, some of which were owned by the government or the nobility, were turned into collectives or state-run facilities. More experiments and diversification of wine ensued, particularly through research at the Magarach Viticulture Institute in Yalta and at the champagne factory in Novy Svyet. After irrigation was brought via the North Crimean Canal, grape cultivation expanded into the plains. Thus, the area under grapevine cultivation in Crimea rose considerably: from 11,800 ha in 1940 to 112,600 in 1970, although problems with infestations of *Phylloxera* lice in the 1970s reduced the area to 86,200 in 1980 (Rybintsev and Petkilyova 1992).

A period of decline occurred in the late 1980s and early 1990s as Mikhail Gorbachev's alcohol prohibition campaign led to the systematic destruction of vineyards. Once prohibition was lifted, however, viticulture began to thrive

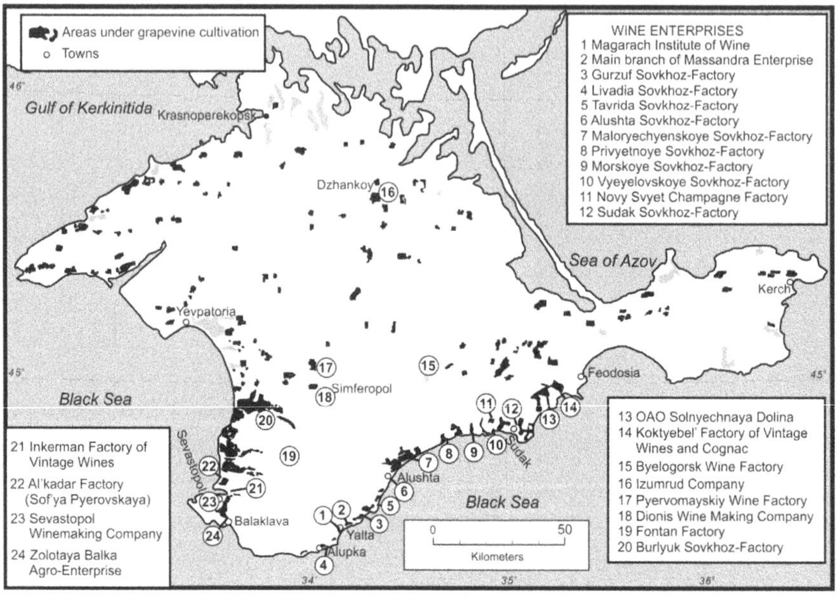

Figure 8.2: Map of current grape cultivation plots and wine enterprises. Map composed by the author using maps, images and a number of other sources.

once more. Grape varieties survived in areas where they had been used for experimentation and research. However, the prohibition campaign was a tremendous blow to the Crimean economy—land under grapevine cultivation in 1993 was estimated at 67,000 ha (Rybintsev 1995), almost the half of what it used to be in the 1970s.

Today, areas under grape cultivation cover most of the peninsula, with the exception of the mountains (Figure 8.2). There are 56 wine and 42 table grape varieties of commercial importance produced in Crimea, some of which correspond to climatic and soil characteristics of certain regions (Rybintsev 1995). The south coast is better known for dessert and fortified wines, particularly sweet ones like Muscat or Sherry (locally known as Xerez). The eastern part of the south coast remains the traditional area for champagne (Novy Svyet and Sudak) and cognac (Koktyebyel'), although many other wines are also produced (Figure 8.2). The western piedmont is better known for various types of table wine and for semi-sweet varieties. The eastern piedmont is known for varieties for sparkling wine and grapejuice. The plains have several varieties: the north produces mainly table grapes, the center produces grapes mainly for dessert and table wines, whilst the south and west produce mainly for grapejuice (Rybintsev 1995). Many of the industries are located on the south coast—literally the birthplace of the Crimean viticulture industry—but others are based in major cities elsewhere (Figure 8.2). A sector of the viticultural industry is merging with tourism, to create oenological tourism, which is now attracting more and more foreign visitors.

The North Crimean Canal and the end of the steppe

Despite their relatively fertile soils (several types of chernozem and kastanozem on loess), the plains were difficult to farm due to low and variable rainfall. Interestingly, a relatively moist phase at the beginning of the Graeco-Roman period permitted Greeks and Hellenized Scythians to farm large areas of the plains (Figure 7.1). But, since then, dry climatic conditions have kept the area only as grazing ground. After the Russian annexation, attempts to develop the rich chernozem soils of the steppe failed due to a series of droughts, particularly in the years 1794, 1799 and 1800, and then again in 1833 and 1837 (Dyulichyev 1997). Evidently rain was available in substantial amounts in some years, giving the appearance of high prospects for farming, but the high variability in precipitation also meant that there were years of extreme drought and crop loss.

Ideas for building a canal to convey water from the mainland into the Crimean plains have existed for some time. Christian Steven, the botanist who became the first director of the Nikitsky Botanical Gardens, was the first to suggest the idea, in the 1840s (Yena, Yena and Yena 2007). Afterwards, plans for constructing the canal were discussed by scientists and politicians but the project never materialized. Stalin revived the idea but its realization did not take place until after his death, under the aegis of Nikita Khrushchev's Virgin Lands Reclamation Project. Groundbreaking for the canal began in 1961, and the first water began to flow into the northwest part of the plains in 1963 and into the Dzhankoy area in 1965. A second phase of the canal brought water farther into the southeastern plains and the Kerch Peninsula in 1971 (Dogushev 1979).

The Northern Crimean Canal (NCC) branches out from a main canal from the Khratova Dam on the Dnieper River. With a series of pumps, the water is carried south into the Perekop, where it enters the peninsula. Then, near Krasnoperekopsk, the water of the main canal is diverted to different sectors of the steppe via other canals (the Krasnosnamyetsk, Chyornomorskoye, Azovskoye and Raznolnoye) (Figure 1.2). To flow across the uneven landscape, water has to be raised by a series of pumps, which means a high cost in terms of energy (Hellegers and Perry 2006; Pavlov et al. 2006). The total length of the canal is 369 km, plus 36 km of pressure pipe that conveys drinking water to the city of Kerch, and its maximum discharge is 330 m^3 per second with mean annual discharge of 4,200 million m^3 (Pavlov et al. 2006).

The implementation of irrigation through the NCC resulted in an increase in the area of arable land to 380,000 ha in the initial phase of the canal, rising to a current figure of 483,000 ha of plowed land, which represents 69 percent of the total crop land in Crimea (Yakovlyev 2009). This irrigated land is devoted, for the most part, to cereal production (70.5 percent of the total cultivated area), which amounts to about 2,000 tons a year, mainly wheat and, to a lesser extent, rice (Yakovlyev 2009).

In addition to irrigating a vast area in the plains, the NCC waters are used for leaching salts from soils, as long as good drainage (i.e., low water table) exists. The canal also provides water for cities such as Dzhankoy, Kerch and Sevastopol, and for recreation activities along its course. The economic benefits of the canal, however, are offset by a series of problems concerning the environment and agriculture itself. In terms of its environmental impact, loss of habitat ranks as the most important followed by invasion of alien species (fish and shells) via the canal waters (Freshwater Ecoregions of the World 2013). In addition to loss of water through evaporation, the high amount of energy needed to operate the pumps represents an input of carbon into the atmosphere.

Of the problems for agriculture itself, the canal has caused secondary salinization, produced by water leaking from the canal and feeding into the water table, which in the northern part of the region is naturally high (see Chapter 9). The accumulation of salts impedes the development of crops, resulting in the abandonment of land. Furthermore, the high cost of maintaining the canal infrastructure and the operation of the pumps makes the cost of agriculture high. During the Soviet period, the state absorbed the high expenditure but, with the recent process in land privatization, farmers are having difficulties meeting the costs (Pavlov et al. 2006). This problem, in tandem with secondary salinization, has led to the idleness of some land.

The closure of the canal that feeds the NCC by Ukraine, following the annexation of Crimea by Russia in March 2014, has reduced the flow of water into Crimea to the point that many of the canals have dried out. However, recently water from dams in the mountains has been diverted to the canal network to irrigate crops. Although local newspapers speculate about deep drilling to obtain underground water, or the building of more dams in the mountains, it is unclear if irrigation agriculture will change in the future. It is also possible that an arrangement between the two countries could be reached and water can be again delivered to Crimea. But at this point it is difficult to assert what the future developments of agriculture in the plains will be.

The Tatar population and the landscape in recent times

The demographic changes that the Tatar population underwent after the late 1700s also meant changes in the cultural landscape of Crimea, particularly through Tatar emigration and the subsequent establishment of a population of Slavic origin. This process has occurred in three phases. The first phase took place during the Ottoman–Russian wars and annexation, when between 20,000 and 30,000 emigrants left, deserting villages and depopulating cities (Fisher 1978). A second phase occurred after the Crimean War and the Ottoman–Russian War of 1877–8. As a consequence, 131,000 Tatars left by 1863, and an additional 30,000 by the early 1890s (Dyulichyev 1997). This emigration considerably reduced the Tatar rural population, resulting in the partial or total abandonment of 687 rural settlements (Dyulichyev 1997). A third phase occurred during the twentieth century, first in the 1920s when many Tatars left or were forced to migrate during the Russian Civil War, and second in 1944 when almost the entire Tatar population was deported to Central Asia after being accused of collaboration with the Nazi occupants (Fisher 1978; Williams 2001).

Despite large waves of emigration, many of the social structures of the Tatar population survived through the Russian Imperial period. Political and economic changes during the early years of the Bolshevik Revolution, and particularly collectivization, eliminated many of the social relations, particularly those linked to religious and feudal structures (Fisher 1978; Williams 2001). The mass deportation of 1944 was the greatest blow to the Tatar cultural presence in Crimea and its subsequent replacement by the Slavic population.

Under Mikhail Gorbachev's reformist policies, the exiled Tatar population was allowed to come back to Crimea. The returning Tatars settled predominantly in areas of the piedmont and in some of the interior valleys of the mountains, particularly the Baydar Valley (Williams 2001). In these regions the cultural landscapes associated with the returning exiles are characterized by settlements with two- to four-storey houses and courtyards surrounded by tall fences. Bakhshisaray, the former Crimean Tatar khanate, is becoming the cultural center of this new revival of Tatar nationalism.

One of the main legacies of Tatar-Turkic culture is the abundant number of place names in the Crimean landscape. During the Soviet period, and particularly after the exile of the Tatar population in 1944, many Tatar place names were replaced by Soviet or Russian names. But a reversal began in the 1990s, in part by the return of Tatars from exile and in part by the general de-Sovietization of place names that took place throughout the former USSR at the time.

Assessment of recent landscape change

The collapse of the Soviet political system and the independence of Ukraine in 1991 meant deep political and economic changes for Crimea. At independence

Ukraine reaffirmed its possession of Crimea, an action that created tensions with the Russian Federation and the strongly Russianized sector of Crimea's population. The main tensions focused on the possession of the naval bases. The two countries made agreements regarding the lease of the Russian naval bases, even though tensions have re-surfaced now and then. Among other changes that came with independence was the shift of central power from Moscow to Kiev, which also meant political and cultural changes. In 1991 Crimea was given the status of autonomous republic, while the Sevastopol district was created as a separate municipality answering directly to Kiev. Political concern regarding the ethnic Russian population, a majority in Crimea, and the increasing numbers of the returnee Tatar population, added to tensions in the country's politics.

The collapse of the Soviet Union created a tremendous economic crisis, not only in Ukraine but in all of the former Soviet republics. Industry suffered a tremendous impact not only because of the shortages of energy that characterized Ukraine in the 1990s, but also because of the disruption to markets caused by the political changes. The political changes of the early 1990s affected agriculture for the reasons discussed above concerning irrigation, as well as the need to adapt to the new political and economic roles. The liberalization of economic policies, including the decentralization of the economy and privatization, meant changes in the landscape, particularly in the rapid development of settled areas and the proliferation of small industries.

Among recent changes, urban expansion is perhaps the most obvious, and the most threatening to the environment. Urban areas have expanded to an unprecedented extent as more people have built houses outside the urban perimeter in areas that were previously devoted to agriculture, pastoralism, military training or simply forests and parklands. Other areas have also seen the impact of urban development, particularly around Simferopol and along the south coast. This rapid expansion has threatened ecosystems and habitats of endangered species, as well as the general value of the natural and cultural heritage of the peninsula. The damage caused by development and the efforts to curb it are the topic of the next two chapters.

CHAPTER 9

HUMAN IMPACTS ON LAND AND SEA

Approaches to landscape transformation

Although environmental history is a young discipline, dating back only to the 1960s, a great number of approaches have developed within it. These approaches depend on the particular academic background of the researcher, the subject of the research, the timeframe and the region. Although strongly dominated by historians, environmental history as a discipline also includes the fields of geography, sociology, archaeology, and political science, among others (Hughes 2005).

The influence of geography in the shaping of environmental history is undeniable, particularly when the subject is regional. Thus, despite being part of eastern Europe, the former Soviet Union and Ukraine, and now Russia, Crimea has an environmental history defined first and foremost by its geographic location. Crimea, unlike other parts of Ukraine and the former Soviet and Russian Empire, is located at crossroads between the vast Eurasian mainland and the Greater Mediterranean, of which the Black Sea is allegedly part. In this respect, the Crimean Peninsula is geographically divided, with its northern part more akin to Eurasia and its southern part to the Mediterranean. In other terms, the Crimean Peninsula is located in a transitional region. From an historical point of view, this transitional situation is even more pronounced: while Crimea belonged to Mediterranean Empires (e.g., Roman, Byzantine and Ottoman), it also belonged to the Eurasian Empires of Tsarist Russia and the Soviets. Thus, an environmental historical approach should be sensitive to both its Mediterranean past and its Eurasian past.

Works on Eurasian environmental history published in the West include a number of monographs and compilations, such as *Studies in Russian Historical Geography* (Bater and French 1983), *East Central Europe and the Former Soviet Union: Environment and Society* (Turnock 2001) and, more recently,

Environmental History of Russia (Josephson 2013). Works on Mediterranean environmental history include such books as *The Mediterranean in the Age of Philip II* (Braudel 1971), *The Mountains of the Mediterranean* (McNeill 1992) and *The Mediterranean: An Environmental History* (Hughes 2005), among many others. However, in addition to using both a Mediterranean and Eurasian focus, one has to be aware that a great deal of the information used for the writing of this book has been obtained from sources published by Russian and Ukrainian scholars who were educated in the German–Russian academic tradition and, later, in the Soviet academic tradition. These works, published in a number of volumes and journals, make up for a good part of the references used in this book.

Although drawing on different approaches, the approach used in the environmental history of Crimea has several distinctive points of its own. First, it is grounded in a strong natural history background (Chapters 3 and 4). Second, it assumes environmental history to have begun with the presence of humans in the Paleolithic, the Neolithic and subsequent periods throughout the industrial era (Chapters 5, 6, 7 and 8). Third, in its evaluation of human impact on the land and surrounding seas it opens up a way to evaluate environmental awareness and conservation in the context of geographic resources such as climate, geology and flora (Chapters 9, 10, 11 and 12).

Land degradation: A historical geographic perspective

One historical irony in Crimean environmental history is the fact that at present the most degraded landscapes are in the plains region, which is precisely the region in the peninsula that historically received the least human impact. The transformation of the plains occurs basically in the late twentieth century. The Khrushchev era, with its Virgin Lands Campaign and the idea of a deep transformation of the landscape for the sake of economic development, has resulted in the strongest impact ever on Crimea's environment, a development that parallels a number of other regions in the former USSR (Josephson 2013).

The impact of historical land degradation in Crimea is difficult to assess as a cumulative process, except perhaps for the past two centuries. Previously, cultures boomed and collapsed; booming populations used the land and then abandoned it, as with many of the cases detailed in Chapter 7. The first Neolithic pulse seems to have died for a while during the drier period evidenced in pollen records around 4,000 BP (Cordova and Lehman 2005). Similarly, the flourishing of the Greek agricultural settlements of the fifth to third centuries BC collapses and abandonment of many lands ensued to the point that even chernozem soils developed on ruined settlements (Cordova and Lehman 2003; Lisetskii and Ergina 2010; Cordova et al. 2011; Lisetskii et al. 2012). Even more stunning is the case of the dense settlement established in the southwestern part of the mountains during the Early Middle Ages (Figure 7.2A). Today the area is covered by dense forest under which one often encounters foundations of walls of ancient farms and villages, events that only recently have been addressed (Schreg 2010, 2011).

As the cases above show, the resilience of the steppe and forests has proved effective in concealing evidence of past population booms. But in more recent centuries, the situation is much different. Areas that were abandoned by Tatars fleeing Russian domination in the late eighteenth century and on many occasions in the nineteenth century were soon resettled with Slavic immigrants.

Whether by natural growth or by immigration, the Crimean population has been growing steadily since the second half of the nineteenth century. With this growth, economic development has also increased steadily, currently raising major concerns for the environment. This recent transformation of the landscape deserves an urgent analysis in order to understand the needs for conservation, a topic that is discussed in the following chapters.

Current environmental issues

Despite the years of stagnation following the collapse of the Soviet system, with privatization and other economic changes, land-use changes became more evident, particularly in the first decade of this century. Despite the slowing of the economy during the post-Soviet crisis, expansion of arable lands shows a steady rise (Table 9.1). In part this increase is due to the slow process of privatization (Hellegers and Perry 2006).

Another sector of land-use that is on the rise is the built environment (Table 9.1). Rapid urban sprawl is evident in almost all Crimean cities. In areas such as Yalta, Alushta and Yevpatoria, the growth of the tourist industry has also had an impact on the expansion of built areas. New settlements have appeared in many areas of the piedmont between Sevastopol and Bakhshisaray. In other cases, suburbs have exceeded the limits of the cities. For example, dachas have sprung up beyond the metropolitan area of Sevastopol, resulting in the destruction of savannoid vegetation, phrygana and shiblyak, as well as remains of ancient Greek farms in many parts of the Heraklean Peninsula (Trelogan 2001). During the Soviet period there were restrictions on settlements near the military bases around Sevastopol, but nine years on from the collapse of the Soviet Union, the number of dachas already covered at least 30 percent of the land unoccupied before 1991.

Growth in certain sectors of the economy, particularly the wine and tourism industries, in tandem with land management malpractices, some

Table 9.1: Changes in area of land resources 2002–8. Area in 1,000 ha per year.

Type of land use	2002	2003	2004	2005	2006	2007	2008
Total	2,608.1	2,608.1	2,608.1	2,608.1	2,608.1	2,608.1	2,608.1
Agricultural land total	1,801.8	1,801.8	1,802.0	1,800.9	1,800.0	1,858.4	1,859.0
Arable land	1,250.7	1,255.4	1,261.0	1,262.2	1,262.7	1,264.06	1,262.0
Monoculture crops	89.4	85.6	83.5	81.9	81.08	80.06	80.06
Fallow land	13.6	14.3	12.6	13.1	13.1	12.06	12.7
Hayfields	2.1	2.1	2.1	2.1	2.1	2.0	2.1
Pasture	446.0	444.4	442.8	441.6	440.3	439.1	440.02
Forests and tree plantations	298.7	298.7	298.6	298.7	299.0	299.2	299.2
Built-up land	102.7	102.9	103.4	107.6	108.1	107.06	107.06
Open, wetland areas	4.9	4.9	4.9	4.9	5.1	5.1	5.1
Open, non-wetland areas	124.6	125.7	129.9	125.8	126.4	126.6	126.6
Water bodies	216.7	215.4	215.4	211.4	211.0	211.1	211.1

Source: Pozachyenyuk (2009a: 642).

inherited from the past and some new, represent issues of great concern for the environment and human health. These concerns include the rapid degradation of vegetation and soils, and the pollution of air, inland waters and the seas.

Forest degradation and wildfires

Quantification of deforestation in the past, particularly before the late 1800s, is difficult because there are no accurate records other than pollen data, references in the literature and, at best, assumptions based on population growth and agricultural expansion. Evidence of deforestation in pollen diagrams points to Greek farming in southwestern Crimea as responsible for the reduction of arboreal species (Cordova and Lehman 2003, 2005). A similar record comes from the Tarkhankut Peninsula, where claims of former woody vegetation have been made (Lyevkovskaya 1970; Hannestad 2007). In the Early Middle Ages a population boom had an impact on the forests of the piedmont and mountains (Schreg 2010). Subsequently, references to timber exports from Crimea by the Genoese provide information on deforestation in the mountains (Yena, Yena and Yena 2004). Information on deforestation of the western end of the mountains as a result of the provision of timber for shipbuilding in the early years of the establishment of the naval base in Sevastopol is better known (Radchenko and Aleyev 2000).

More complete records of forest exploitation exist only from the late 1800s. Since that time, the period between 1888 and 1921 exhibits one of the greatest reductions in forest area. In 1888 the total area of forests in Crimea was 315,000 ha, whilst in 1921 the total area of cover had been reduced to 243,700 ha, a reduction of 23 percent in 33 years (Mishnyov 2005). Although stricter laws established in subsequent years have reduced the rate of forest diminution, forest destruction has continued at times, particularly during World War II, when fuel shortages required the use of forests. Additionally, heavy fighting destroyed some forests areas, particularly around Sevastopol, as is shown in some photographs. Nonetheless, today most of these areas again have once more seen forest re-growth. The boom in state-sponsored tourism during the postwar years has also resulted in some reduction of forests through the construction of camps and accommodation facilities. But despite these setbacks protection has somehow been effective, so that in many areas forests have regenerated. Nonetheless, threats in the form of fire, illegal cutting and urban growth still exist.

Forestry management in the Autonomous Republic of Crimea is the responsibility of the Republican Committee for Forestry and Hunting (RCFH). The objectives of this organization are multiple, most notably the protection and rational use of forests, the prevention of fires, fighting fires, reforestation, and, as its name indicates, the regulation of hunting. To meet its objectives, the RCFH is divided into 15 regional offices, some of which include areas of the plains without proper forests. But most of the forests in the mountainous south of the peninsula fall under the jurisdiction of regional RCFH offices, including those of the Crimean Mountain Preserve.

Wildfires rank as the largest threat faced by the Crimean forests. Forest fires are strongly linked to tourism, particularly careless campers, who occur in considerable numbers during the hottest and often driest times of the year. This threat has been one reason why many sections of the forest have been closed to tourists during the hottest and driest days of the summer. Next to the fire problem is illegal cutting, which is equally difficult to control despite efforts at law enforcement. Finally, threats to forests also include the expansion of urban centers and tourist complexes, which are already putting pressure on the edges of most forests.

Degradation of non-forest communities

Of the ecosystems outside forested areas in Crimea, the steppe is perhaps the most important. Non-forest communities include the steppes of the plains and the meadow, petrophytic steppe of the yaila, the wetlands ecosystems in coastal areas and, to a certain extent, the phrygana and shiblyak, the two main communities of the Mediterranean shrubland of the south coast. The latter are included if we assume that they are secondary communities produced by the degradation of forests. (However, see the discussion regarding the shiblyak in Chapter 12.)

Afforestation is one of the main threats to the steppe of the plains and the non-forest areas of the yailas. The origins of this issue are highly complex but, overall, they derive from the general forest-oriented conception of land reclamation and conservation that has prevailed in the country to this day. Although many activists in Crimea and Ukraine in general blame this conception on Soviet conservation policies, the problem seems to go back to the nineteenth century, as suggested by numerous government documents (Moon 2013). Throughout the twentieth century, many times the solution to the problem of land degradation, particularly soil erosion, has simply been to plant trees.

Unlike reforestation, which means re-planting trees in areas formerly covered by forests, afforestation is a more generic term to refer to tree planting anywhere, whether it suits nature conservation or not. In terms of ecosystem restoration and conservation, afforestation is not suited to the steppe of the plains and the meadows and petrophytic steppe of the yailas, where no evidence of former forests exists, at least during the Holocene. The problem is evident from the very fact that most of the planted trees on the steppe of the plains and the yailas do not survive the dryness, soil salinity and strong winds. Those that do survive create competition with the native herbaceous vegetation, often driving it into endangerment. The topic, however, is discussed in more detail—especially its biogeographic and conservation impact—in Chapters 10 and 11.

The steppe of the plains is the most affected of all the vegetation regions. Unplowed and undisturbed steppe in Crimea accounts for only 10 percent of the original steppe (Vakhrushyeva 2009). As described in Chapter 8, the destruction of the steppe ecosystems reached its peak as a consequence of the Virgin Lands Campaign of the 1950s and 1960s, a project that materialized with the possibilities of irrigation brought by the construction of the North Crimean Canal.

The steppe that was spared by the Virgin Lands Campaign, however, now faces many threats, particularly pollution, invasive species and afforestation. Pollution of soil, water and air has been singled out as the main threat in the steppe around the Sivash (Baykov 2009). Soil pollution usually comes with irrigation and mining, which often increases minerals in the soils and secondary salinization. The pollution of water, in particular, endangers many of the already threatened amphibians and birds of the region. Invasive species are another problem that is linked directly to agricultural activities. Many of the invasive species are non-native and ruderal (i.e., weeds). They originate on fallow land or along canals, from where they jump over to the nearby steppe, where they compete with the native species, some of which are already endangered.

The meadows and petrophytic steppe of the yailas face a series of threats, of which the most significant are the effects of irresponsible tourists and afforestation. Driving cross-country, a common occurrence in some of the yailas, leads to the destruction of herbaceous communities, soil compaction and, in some cases, soil erosion. Afforestation, as in the case of the steppe, means competition with the native herbaceous vegetation and an increased risk of fires.

Wetland environments in Crimea are located mainly along the coast and in the areas around the Sivash and south of the Perekop. Like their steppe counterparts, wetlands are threatened by water pollution, particularly from mining in areas near the Perekop, or by oil and waste water near the ports, particularly in the bays around Sevastopol and Kerch (Baykov 2009).

Overall, human impact on non-forest areas has caused a considerable reduction in biota, particularly in the steppe and wetlands, where the number of fauna has declined considerably (Kryukova et al. 1988; Biodiversity Support Program 1999). Many floristic species have also been threatened, including some of the grasses that were once abundant in the steppe and other non-forest areas of Crimea: e.g., the 13 grass species of the genus *Stipa* are on the list of endangered species in the *Red Data Book of Ukraine*. *Stipa* grasses were once an edificator and co-edificator of most of the steppe vegetation types of Crimea, as names such as "*Stipa-Festuca*" or "*Festuca-Stipa* and mixed herb steppe" indicate (see Figure 4.3).

Soil degradation

The various environments and the diversity of soils in Crimea present different susceptibilities to human and non-human impact. Thus, land degradation varies depending on the soil and the type of impact: dehumification, deflation, erosion, karstification and salinization (Dragan 2004). Many of these forms of degradation, although carried out by natural processes such as wind and water, are catalyzed by human disturbance.

Dehumification and deflation are problems directly related to agriculture. Dehumification is a form of degradation strongly associated with intense irrigation, as water percolating through the soil washes organics and nutrients to lower levels. Deflation is a common factor that leads to the loss of fine material, nutrients and organic matter by wind. It happens when the natural vegetation cover is removed and during times when the soil has no crop.

Salinization is a natural process, but its human-induced form is secondary salinization, a problem that results fundamentally through irrigation malpractice. Of the problems for agriculture itself, the canal network of the North Crimean Canal has caused secondary salinization. This process is produced by water leaking from the canal and feeding into the water table, which in the northern part of the region is naturally high (Mikhaylov 2009a). As the water table rises close to the surface, high evaporation rates induce precipitation of minerals, particularly sodium and sodium-based minerals (e.g., salts), which prevent the growth of plants with little or no tolerance to salinization.

Soil erosion occurs in different forms and for a variety of various reasons. The problem of soil erosion seems to have a millennial history in Crimea, probably dating back to the Neolithic period (Cordova and Lehman 2012), but definitely with the advent of Greek farming (Lisetskii et al. 2012). As in the Mediterranean region, however, the topic of soil erosion and soil resilience is still controversial given the long-term nature of the process, sometimes making it difficult to discern human from natural causes. But the general consensus is that the human factor, through activities such as agriculture and pastoralism, has been the most important, particularly in the Holocene (see, for example, cases presented by S. Bottema, G. Entjes-Nieborg and W. van Zeist 1990). Soil and pollen records in the lowlands (e.g., the Heraklean Peninsula, the Chyornaya Valley and the Kerch Peninsula) show that farming and pastoralism had a strong influence on soil erosion and vegetation transformation (Cordova and Lehman 2005; Cordova et al. 2011). The case of the Crimean highlands, particularly the yaila region, shows even more clearly post-Neolithic erosion of slope soil led probably by pastoralism (see Chapter 6). Soil erosion continues to this day. The areas with the strongest rate of soil erosion are along the south coast, particularly on the Jurassic flysch deposits, the Heraklean Peninsula, large tracts of lands in the Tarkhankut and Kerch peninsulas, and on the piedmont, where in most cases trails, roads, removal of vegetation and construction of structures contribute to the destabilization of slopes.

The ever-growing problem with water

The Crimean Peninsula has one of the lowest water-supply potentials in Ukraine. In terms of per capita water supply potential, each Crimean resident has five times less water than each resident in the rest of the country (Mishnyov 2005). The problem of water supply and water use in Crimea has two sides: a natural side and a human one. The natural side relates, on the one hand, to the relatively small catchments in the peninsula, and on the other hand, to the uneven distribution of running and underground water as a result of the topography and geology of the peninsula. Most of the rock structures of the Crimean Mountains dip northwards (Figure 3.2): therefore, the structures that transport water from the karstic system of the Middle-Upper Jurassic to the aquifers and springs benefit the northern slopes. Furthermore, the largest basins are on the north side of the mountains, meaning that the catchment of rainwater favors the rivers that drain the northern slopes. Conversely, the southern slopes of the mountains get less groundwater and run-off: their river basins are too small to catch and carry enough water for storage in reservoirs or, in general, to supply water to residents and agriculture. But despite the hydrological benefits of the river basins and aquifers of the north slope of the mountains, the amount of water available is not nearly enough to supply water to the growing population, let alone to provide irrigation for crops. Moreover, the plains, with lower precipitation and relatively deep younger sediments, have low prospects in terms of hydrological resources.

The problem of water supply has been addressed by the construction of dams. This in turn has favored the growth of towns and cities on the piedmont and even on the south coast. For the latter, in particular, the problem was tackled in the early 1960s when water began to be diverted from upper Bel'bek River into Yalta through a 7,125 m long tunnel through the Main Ridge of the mountains. The damming of streams and the creation of reservoirs, however, has not been enough to satisfy the high water demand—one reason for the construction of the North Crimean Canal. In the end, the construction of dams and canals not only meant disruption of aquatic fauna but also the destruction of most of the natural vegetation of the plains. As mentioned in Chapter 8, with the closure in April 2014 of the canal that feeds into the North Crimean Canal from the Dnieper, agriculture is relying more on water from the dams in the mountains. It is possible, then, that this could lead to more water damming and deeper transformation of habitats along the Crimean rivers.

Air and inland water pollution

The worst areas of air pollution in Crimea are the area of the plains south of the Perekop, the Kerch, and the districts of Simferopol, Bakhshisaray and Sevastopol, where most of the processing and manufacturing plants are located. Pollution in the Perekop area is related to the mineral and salt extraction industry. The pollutants that exceed acceptable levels in this area are nitrogen dioxide, carbon monoxide and hydrogen fluoride, and dust particles, among others (Baykov 2009).

In the Kerch area the pollution is associated with the fish processing plants and port activities, including the pollution produced by passing ships. Economic decline and the closure of some factories, however, has resulted in a decrease in pollutants, from 10,650 tons per year in 2000 to 1,777 tons per year in 2008 (Baykov 2009). Other polluted areas include the Simferopol, Bakhshisaray and Sevastopol districts, where some decline has also been recorded. Despite the decline, pollution levels are still too high, and in some cases the levels are higher than the permissible maxima (Baykov 2009). Although the majority of polluting industries and agriculture are located on the piedmont and the plains, with very few in the south coast region, the latter is not spared from pollution. Heavy traffic along the main coastal highway contributes to a tremendous increase in smog, which in turn is trapped between the sea and the mountains (Baykov 2012).

One general way of assessing river pollution in Crimea is to compare it with other regions of Ukraine. In general, pollution concentrations in Crimea are less than in most regions of Ukraine but, because industries in other regions of Ukraine have been in much more rapid decline, the rate of annual increase of pollution concentration is lower. In the period 1993–2000 water pollution in Crimea increased 210 percent more than the average concentration of pollutants (Nazarov, Cook and Woodgate 2004). In contrast, the rivers of Northern Donetsk—the most polluted region of mainland Ukraine—during the same period saw only 92 percent increase above the average concentration of pollutants (Nazarov, Cook and Woodgate 2004). But it is also important to acknowledge that the types of pollutants dumped in the Crimean streams are different from those in other regions of Ukraine, where coal mining and chemical industries are located. Most of the river pollution in Crimea comes from agriculture and waste water; the latter in particular is the result of improperly treated water being dumped into the rivers, leaks from the main sewage systems, and landfill sites (Nazarov, Cook and Woodgate 2004).

All Crimea's rivers are polluted and all exceed the maximum admissible concentrations of pollution, with the exception of those streams in catchments very high in the mountains. Pollution is not only related to sites with more industries and agriculture: most rivers emptying on the south coast, particularly those in the Yalta and Alushta areas, are highly polluted, particularly with wastewater, something that has become a real problem for bathers in the coastal resorts (Byeryegovaya 2011).

The main source of pollutants in the plains region is irrigation agriculture, salt mining, and pollution leaking from vehicles along the roads into ditches and fields (Baykov 2009). The impact is detrimental not only to the flora and fauna of the wetlands, the soils and the surrounding seas, but also to humans who have direct contact with the polluted waters.

Black Sea pollution

To assess aspects of sea water pollution around Crimea, it is important to evaluate the overall environmental issues of the Black Sea Basin, which for the purposes discussed here include the Sea of Azov. The most common environmental issue is pollution in all its forms, chemical (organic and inorganic) and biological. The landlocked nature of the Black Sea Basin makes pollution problems worse, given the low rate of self-purification (Zonn, Fashchuk and Ryabinin 2008). Additionally, the largest rivers emptying into the Black Sea drain the most developed agricultural and industrial areas of eastern Europe: 80 percent of the run-off into the Black Sea comes from the Danube, Dnieper and Dniester. The Danube has a water input of 200 km^3/year, the Dnieper, 50 km^3/year; and the Dniester, 10 km^3/year (Kosarev and Kostianoy 2008).

The total input of water into the Black Sea is approximately 350 km/year from run-off, 225 km^3/year from precipitation, and 175 km^3/year from the influx of sea water from the Mediterranean via the Bosporus, while the losses are 350 km^3/year from evaporation and 225 km^3/year from outflow through the Bosporus into the Mediterranean (Kudelsky 2011). The loss by evaporation is twice the input precipitation, and the loss through the Bosporus is about one-third more than what flows in through the same strait. These dynamics result in an increase in salinization (average 19 percent), which on average has been increasing since the second half of the twentieth century given that the damming of rivers has reduced the amount of fresh water coming to the sea (Kudelsky 2011). The impounding of the Dniester, Dnieper and Don rivers has reduced the flow of water into the Black Sea and Azov Sea by 50 percent (Josephson 2013).

The increased salinity in the Black Sea affects sensitive marine and estuarine species. This problem has become even greater in the Azov Sea, where natural salinity is low enough to allow riverine fish species to live. The damming of the Don River caused not only a decline in the freshwater flowing into the Azov Sea, but also a slight drop in the sea level that has been balanced by the flow of more saline waters from the Black Sea (Kosarev and Kostianoy 2008; Josephson 2013). As a consequence, the number of fisheries declined so dramatically in the 1960s that a project to build a dam across the Kerch Strait was proposed, though later abandoned (Josephson 2013).

The main types of pollution in the Black Sea include petroleum products, pesticides, organic and mineral pollution, and heavy metal pollution (Zonn, Fashchuk and Ryabinin 2008). Petroleum products are strongly related to shipping, a problem that is heavy around port facilities and along the straits. This type of pollution has increased so dramatically that it has now, for the first time, exceeded that of pesticides (Zonn, Fashchuk and Ryabinin 2008). Organic pollution comes from waste waters and byproducts of erosion and agriculture. The influx of organic compounds from organic waste and fertilizers into the Black Sea has created eutrophication, where the amount of nutrients causes an increase of phytoplankton, which in turn depletes oxygen levels (Kudelsky 2011). In addition to the reduction in oxygen, eutrophication of the Black Sea waters has created an ideal environment for invasive species such as comb jelly (*Memiopsisleidyi*), which have wiped out anchovy among other species (Chashchin 1996). Additionally, it has led to the overproduction of certain algae that produce red tides, which have detrimental effects on fishes (Zonn, Fashchuk and Ryabinin 2008). The problems created by eutrophication have led to a decline in commercial fish species, such as the highly valued sturgeon.

The increase in heavy metals in the waters of the Black Sea is considerable, but still ranks below pesticides. Normally heavy metals—particularly copper, zinc, nickel and cadmium—are distributed unevenly, being more prominent in the sediments near river mouths and along the coast, and affecting mainly communities living on the marine shelf (Zonn, Fashchuk and Ryabinin 2008). Although radioactive pollution exists in the Black Sea, the levels are not particularly high, contrary to popular belief. Even after the Chernobyl accident the maximum concentrations of ^{137}Cs never exceeded 14 pCi/liter of sea water, which is 1,000 times less than the concentration allowable for drinking water (Kudelsky 2011).

Taking all types of pollutants into consideration, the most affected part of the Black Sea is its northwestern part, where the waters of the Danube, Dniester and Dnieper converge (Zonn, Fashchuk and Ryabinin 2008; Kudelsky 2011). The most polluted parts of the seas around Crimea are those near the port facilities, particularly around Kerch and Sevastopol. The Kerch Strait is subjected to high traffic, resulting in constant spills of hydrocarbons and materials discarded by ships. At the other end of the peninsula, Sevastopol Bay has even worse problems: fuel leaking from ships, pollutants from factories, and waste water in the almost closed harbor, make Sevastopol Bay the most polluted water inlet of the Crimean coast. The two most prominent contaminants are hydrocarbons and organic pollutants, both of which are creating not only overproduction of plankton but also reductive conditions that are more usually associated with swamps (Osadchaya et al. 2003). With practically no tides, the exchange of waters between the harbor and the open sea is minimal, a situation that is exacerbated by the defense wall across the entrance of the harbor, which traps the pollution inside (Radchenko and Aleyev 2000; Gordina et al. 2001). Although the wall prevents water exchange, it does not keep pollution from flowing out of the harbor. Additionally, outside the wall, pollution produced in the port facilities in the bays of Stryelyetskaya, Kruglaya and Kamishovaya adds to the pollution problem of the waters off the southwestern Crimean coast.

A major source of pollution in Sevastopol Bay is the navy, which is beyond municipal control. The Russian navy pays compensation for pollution damage to the government of Ukraine, not to the Sevastopol Municipality (Radchenko and Aleyev 2000).

Although fisheries inside the harbor declined a long time ago, the pollution leaving Sevastopol Bayisnow affecting a wider area (Gordina et al. 2001). Health problems for holidaymakers swimming in the highly contaminated waters near the beach resorts of the region have been on the rise, particularly when water temperatures reach around 25°C in the summer (Byeryegovaya 2011).

The pollution related to port facilities is expected to rise in Crimea, as proposals for the creation of new harbor facilities do exist, particularly for the west coast. In 2004 a proposal to construct a mega-port facility in Lake Donuzlav was signed by then-President Viktor Yushchenko. In technical terms, the proposal noted that this is an ideal location for a harbor because it is the deepest of the coastal lakes in Crimea (25 m) with practically no freezing during the winter. In fact, a Soviet naval base has used the lake for large warships. Although Chinese capital was to provide the funds for the port facility, the project never materialized given the general opposition of local inhabitants and environmentalists. Later, even the president himself showed his neutrality to the project. Politicians of the Crimean Autonomous Republic became divided over the issue and the project did not go forward.

After the failure of the mega-port for Lake Donuzlav, another proposal appeared, this time for the Saki district in lands belonging to the village of Frunze. The design envisions the transformation of the area to accommodate docks and container facilities by modifying the coast and completely obliterating Lake Bogayly and impacting nearby Lake Kiziyl-Yar. The port project is expected also to affect local tourism, particularly in nearby Lake Saki, a resort that attracts visitors for the curative properties of its muds, as well as along the nearby beaches which are frequented by bathers in the summer. In addition to the modification of the lakes and the coastline, the projected port is expected to produce large amounts of pollution from the heavy traffic, as well as potential accidents and spills, thus creating practically similar levels of pollution seen in the waters around Sevastopol.

How are the local residents responding to the construction of the port? In contrast to the villages around Lake Donuzlav, the residents of the Frunze village did not oppose the plans. As one resident put it, "We have raised chickens all our life: we can handle ship containers" (Zakirov 2013). Yet, some residents have doubts, and the general public opinion, among both Crimean residents and visitors, shows general opposition to the project (Zakirov 2013). Construction was scheduled to start in 2014 with the technical and financial support from the Chinese company China National Technical and Export Corporation, but with the sudden annexation of Crimea in March 2014, the project failed to materialize, since Ukrainian politicians were directly involved in the project.

What lies ahead?

Compromise between economic development and nature conservation is a new dilemma for a Crimea that is still struggling to escape post-Soviet economic stagnation. Capital investment should boost the economy and create jobs, particularly in areas of agriculture, trade and tourism. This means, however, compromising the environment along with all the endangered species, and even compromising human health. Regulation for sustainable development is possible, but prospects are very low, as it is in other European countries, including some of the former Soviet bloc.

As the pages of this chapter are written, a great struggle is taking place in Ukraine over the country's economic future. For some, the best bet lies with the European Union, for others in strengthening the centuries-long link with Russia. The debate, carried out in the *VerkhovnaRada* (Supreme Parliament) of Ukraine and in the streets of Kiev and other cities, shows the contradictions between being part of Europe or part of the greater East Slavic culture, which is one of the roots of the Euro-Maidan political movements in 2013–14. The debate in itself brings uncertainties as to the economic future of the region and the protection of nature. Under Russian rule things may continue the same or they may change, but it is too early to see any new transformations.

CHAPTER 10

BIOGEOGRAPHIC BACKGROUND TO CONSERVATION

The scientific basis for nature conservation in Crimea relies on studies of a variety of biogeographic topics such as biodiversity, relict populations, rare species and endemisms. These topics have been investigated in the course of the twentieth century by a number of scientists who proposed a series of hypotheses. At the time these hypotheses were difficult to test but recently, in the context of new paradigms, new technology and international cooperation, these they are being reconsidered.

The biogeographic topics discussed in this chapter refer primarily to Crimea's evolutionary and historical biogeography, largely those concerning the origin of Crimean vegetation, relict flora, the number of endemic species, and biodiversity. Additionally, the concept of Mediterraneanization, both physical and cultural, is included here as an important biogeographic aspect of the Crimean environment that deserves discussion because of its links with conservation and natural and cultural resource management, and its significance in tourism development. For the purpose of describing biogeographic phenomena, the physiographic regions of the peninsula are here divided into two: the Crimean plains, and the Greater Crimean Mountain Region (GCMR), which encompasses the mountains proper, the piedmont, and the south coast.

The origin of Crimea's flora

Historical development

Since the works of Hablitz (1799) and Pallas (1812, but first published in Russian in 1795) appeared, Crimean flora has intrigued botanists, who soon began to speculate on its origins, uniqueness and connections with other regions. It was Christian Steven, first director of the Nikitsky Botanical Gardens, who first became interested in the origin of Crimean flora and its endemisms (Yena, Yena and Yena 2007). However, it was not until the early twentieth century that more solid hypotheses about the origin of Crimea's native flora and its endemic species begin to emerge. Of these, the works of Yevgeniy V. Wulff in the 1920s sparked the start of a more systematic focus on the historical biogeography of Crimea's flora.

The historical development of the ideas and discussions concerning the origins of Crimean flora and its endemic and relict species are described and discussed among others by Ye. V. Wulff (Vul'f 1926, 1944), G. E. Grosset (1979a, 1979b), V.P. Malyeyev (1946, 1948a, 1948b) and N. I. Rubtsov (1980). Of these, G.E. Grosset's articles provide a historical review of past works, as well as proposing new ideas, at the time marking a new direction in the historical biogeography of Crimea's flora. More recent works have introduced new ideas and scenarios, including Ya. P. Didukh (1992) and Golubyev (1996b), although the lack of paleobotanical evidence to support their ideas is still an issue. More recently, Andrey V. Yena (2007, 2012) has discussed the history of studies concerning new theories on the origin of Crimea's flora and its endemics.

One feature that is apparent in most ideas and hypotheses concerning the origin of Crimea's flora is the emphasis on Mediterranean floristic affinities. However, the works of V. P. Malyeyev (1946, 1948a) also focus on a number of flora that include boreal relicts as well as elements found today in the steppe. Malyeyev also pioneered the idea of glacial floristic refugia in Crimea. Obviously some of these ideas on both Mediterranean and Boreal relict flora, as well as the issue of floristic *refugia*, can be re-analyzed using the pollen records that have appeared in the past decade, all of which are discussed in Chapters 5 and 6. The Boreal and Mediterranean relict flora result from the fact that Crimea lies on the border of two of the high-rank Palearctic regions: the Circum-Boreal and the Mediterranean (Yena 2012). This location—with the existence of relicts and glacial refugia—makes the peninsula an extremely interesting territory for the study of floristic exchange between the northern and southern Palearctic floras. At a more regional scale, Crimea also straddles the boundaries of the Central Russian province, which covers the northern flat part, and the Ponto-Euxinian province, which cover its mountainous part (Yena 2012). The location of Crimea at the boundary between floristic regions and provinces is complicated by the cyclical cooling and warming of climate, and the changes in sea levels that took place during the Pleistocene. These changes suggest the occurrence of Circum-Boreal floristic assemblages during glacial times, and Mediterranean floristic assemblages during interglacial times. These cycles suggest not only vegetation migration, but also formation of refugia, or areas where floristic species survive under an adverse climate. Although the concept of *refugia* is often used to refer to sheltered areas for thermophilous vegetation during glacial times, it should also include shelters where boreal floristic elements survive interglacial climates (i.e., the Holocene) in middle latitudes.

Relict flora and floristic anomalies are two concepts associated with the origin of native Crimean flora and the concept of *refugia*. Floristic relicts can be of two kinds: geographic and climatic (Huggett 2004). *Geographic relicts* are species that were once widespread geographically but that now have a more restricted spatial distribution, or species that have now been disconnected from the main areal distribution and kin species. The best example in Crimea is *Pinus pityusa*, which is a disjunctive subspecies of the more broadly distributed *Pinus brutia* (Yena, Yena and Yena 2005) (Figure 10.1A). Similarly, *Pistacia mutica*, believed to be a close relative, if not a subspecies of *P. atlantica* (Davis 1982), shows a similar disjunctive distribution, with more outliers in the mountainous parts of east Asia and the Crimean and Novorossiysk coasts (Figure 10.1B). The disjunctive nature of other species, such as *Arbutus andrachne* and *Pyracantha coccinea*, suggests also a strong concentration in southern Crimea and the Novorossiysk area, giving rise to the idea of the species clinging to very small areas with microclimates (Figure 10.2) In contrast, other species of Mediterranean origin, such as *Jasminum fruticans* and *Paliurus spina-christi*, seem to have a less disjunctive distribution, often leaving a trail between Crimea and the Mediterranean region via the west coast of the Black Sea (Figure 10.3). These six examples suggest a number of different ways for the diffusion of these Mediterranean species along the Crimean south coast. Thus, the question of whether these species of Mediterranean origin in Crimea are relicts is still debatable, since they could have arrived in the early Holocene (Cordova 2007).

Climatic relicts are species that were once widely distributed locally, when the climate was different, but which have been cut off from the main distribution area, surviving in small areas. The best example of a climatic relict is the stand of *Betula pendula* near the Babugan Yaila, a locality that in turn should be considered a modern *refugium* of boreal flora. Floristic anomalies are the unusual absence of certain species in a region (Yena 2012). One example of a floristic anomaly in Crimea is the absence of *Castanea sativa*, which is widespread in the Euxinian province, with which the Crimean Mountains have a strong floristic affinity. The causes of floristic anomalies seem to be related to the cyclical changes in sea level associated with glacial–interglacial cycles, which affect the areal distribution of species (Yena 2012).

Floristic relicts and anomalies have been essential for reconstructing Crimea's recent floral history, which throughout most of the twentieth century has strongly been linked to the notion of geological land bridges.

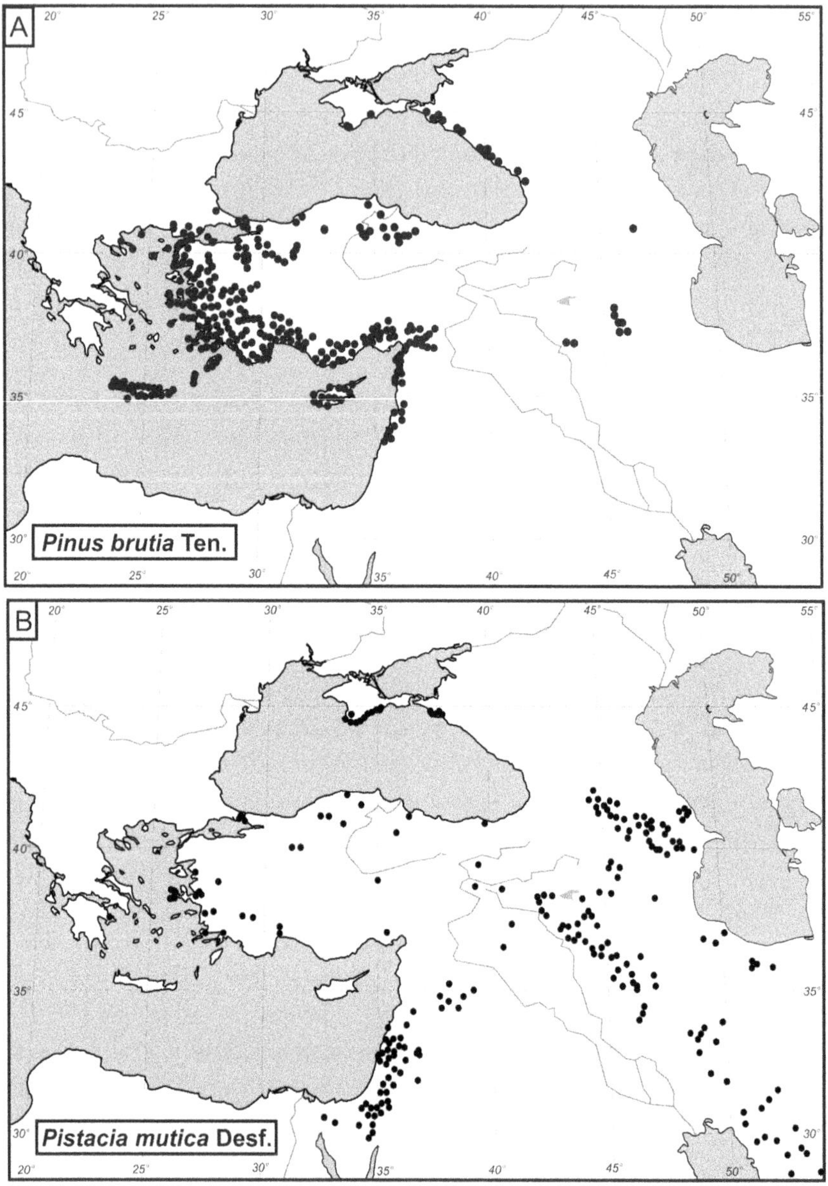

Figure 10.1: Distribution of *Pinus brutia* and *Pistacia atlantica* in the eastern Mediterranean region, the Black Sea Basin and west Asia. Source: Browicz and Zieliński (1982–94).

Today, the origins of Crimean flora are studied under a different paradigm, particularly in the light of new advances in geological, paleoecological and genetic research. Nonetheless, the concept of geological land bridges is an interesting one to review as a stage in the history of Crimean biogeography.

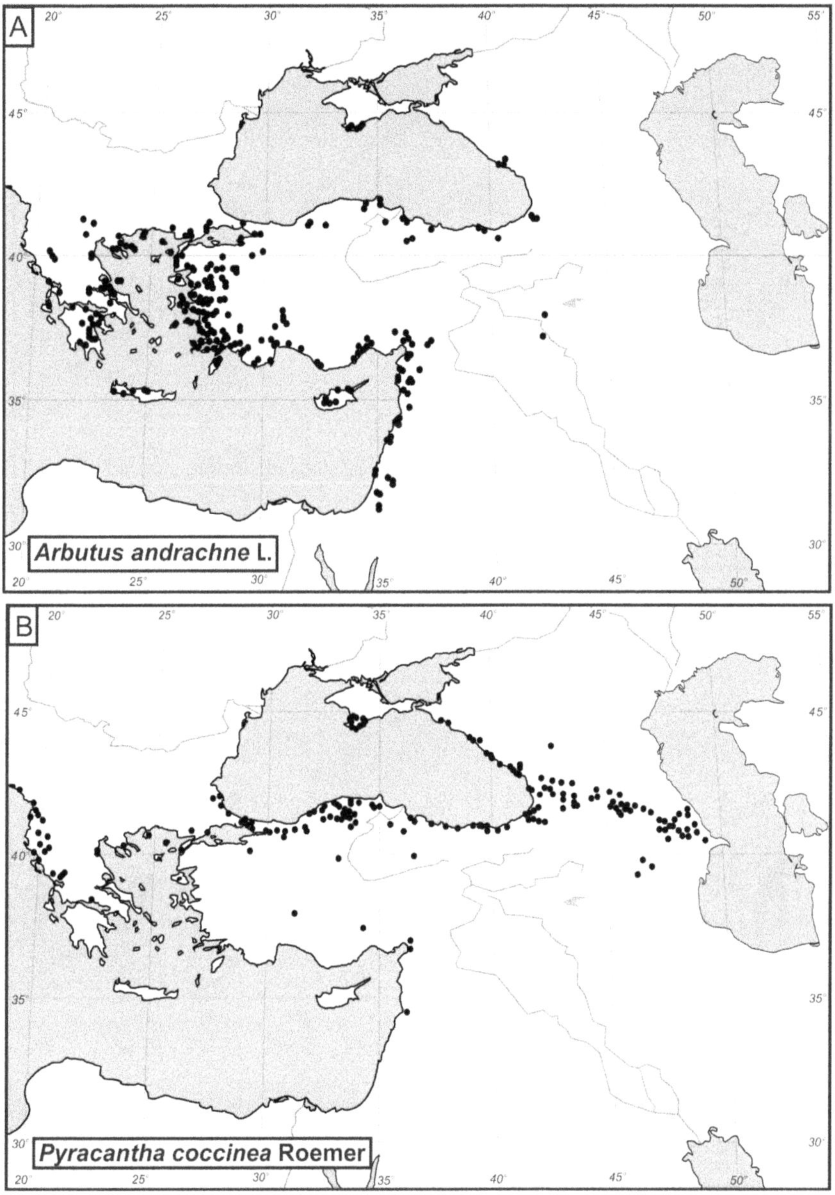

Figure 10.2: Distribution of *Arbutus andrachne* and *Pyracantha coccinea* in the eastern Mediterranean region, the Black Sea Basin, and west Asia. Source: Browicz and Zieliński (1982–94).

The Pontida Hypothesis

Perhaps one of the main aspects of academic interest in the origin of Crimean flora was the strong focus on land bridges, or terrestrial connections with Asia Minor and the Caucasus in the deep geological past—a paradigm that in

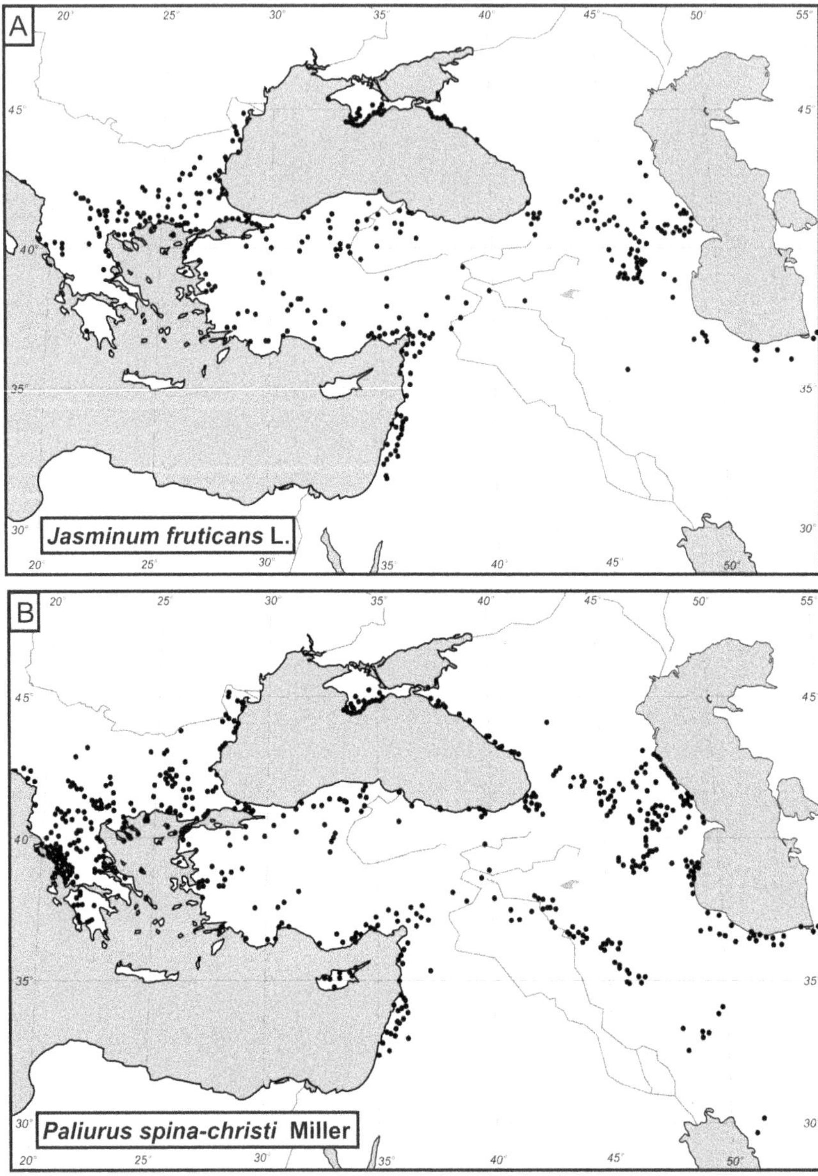

Figure 10.3: Distribution of *Jasminum fruticans* and *Paliurus spina-christi* in the eastern Mediterranean region, the Black Sea Basin, and west Asia. Source: Browicz and Zieliński (1982–94).

Crimean biogeography developed around the Pontida Hypothesis. During the nineteenth and early twentieth centuries, geologists and natural historians believed in the existence of Pontida, a presumed land mass, now under the sea, connecting the Crimean Mountains with Asia Minor, explained the passage of

Mediterranean flora into Crimea (Rubtsov 1980). The idea was rejected by geologists, who presented a different scenario (e.g., Muratov 1974). More recent studies of Black Sea geology have not provided any proof of the existence of such a land mass. Pontida was assumed to have existed in pre-Quaternary times, possibly during the Miocene and Pliocene, at a time when the configuration of the coastline was different. Today biogeographers have realized that much more important in this issue is the coastal configuration changes that occurred in the Pleistocene (Grosset 1979b; Yena, Yena and Yena 2005). However, the idea of Pontida has remained as a symbol of the biogeographic connections between Crimea and the eastern Mediterranean region (Yena, Yena and Yena 2007).

The idea of land bridges, or land connections between Crimea and other regions, has been applied to contemporary Crimean biogeography but within a different framework. Connections with the Caucasus, the Balkans and Asia Minor occurred as large tracts of the continental shelf were exposed when the sea level dropped during the cold stages of the Pleistocene (Grosset 1979b; Yena, Yena and Yena 2004a, 2005; Cordova 2007). With a shoreline 100 m below the present level, the Kerch Peninsula in Crimea and the Taman Peninsula in the western Caucasus were separated only by the Don River, whose mouth was located farther south (Shcherbakov et al. 1977; Fyodorov 1978), and the coast followed a much lower level along the Neo-Euxinian Lake (Figure 5.4). This raises an important question regarding the origin of the highly thermophilic Mediterranean species of the south coast.

Relict flora in Crimea

Mediterranean flora: relict or recent immigrants?

Most species of Mediterranean origin in Crimea are concentrated along the relatively warm south coast, where a number of species have their northernmost distribution (Figures 10.1, 10.2 and 10.3). The question of whether these Mediterranean elements survived the glacial cold in *refugia* or migrated to southern Crimea in the Holocene is one of the main puzzles that remain to be solved in Crimean biogeography. The question is complicated by the lack of continuous paleobotanical records, by issues related to the true endemic species, and by the still unsolved issue of possible refuges on the south coast.

The idea of Mediterranean relict elements from the much warmer climates of the Tertiary period, an idea supported by the Pontida Hypothesis, is not supported by the few palynological records of the Miocene and Pliocene in Crimea—as shown, for example, in the records from Lyubimovka Beach in western Crimea (Parishkura 1978) and various locations on the southern mainland of Ukraine (Shchyekina 1979), both of which fail to show the presence of species linked to the Mediterranean realm. Instead, the prevailing pollen in these sediments points to the presence of taxa such as *Carya*, *Pterocarya*, *Zelkova* and a number of other floristic elements of the Hyrcanian and Euxinian forests of present-day Georgia, northwestern Turkey and northern Iran.

In view of the discrepancies and lack of evidence for an ancient land bridge to explain the presence of Mediterranean flora in southern Crimea, Grosset (1979a, b) put forward another proposal, one that involves changes during the Pleistocene—more specifically, during the Late Pleistocene. He concluded that the main migration paths or land connections existed during the Pleistocene, particularly at times when the sea level dropped, allowing migration through an exposed marine shelf. Additionally, Grosset proposed the transport of seed material via migratory birds. In fact, he believes that the migration of plants from the north, i.e., plants of Eurasian and Boreal affinities, may also be related to seeds spread by birds. A third mode of transportation, he adds, is by early colonists from the Mediterranean region, who may have brought some taxa either deliberately or accidentally.

Grosset's theory of migration along the exposed marine shelf can be summarized in the following scenario for the last glacial–deglacial period. Although the lowest sea level was reached in the coldest part of the last glacial, plants of warmer lands could not migrate north. However, as the climate began to warm up and become moister around 14,500–14,000 years ago, thermophilous species began to migrate north along the still exposed shelf. In reality, the shelf was still mostly exposed around 13,000 years ago, as proposed by more recent studies (see Figure 5.4).

Another means of post-glacial northward migration—the sea currents—was proposed by Cordova (2007). This scenario invokes the establishment during the Holocene of the modern current system of the Black Sea—the west and east gyres (see Figure 1.5). However, Cordova's (2007) scenario implies that it was not until after around 7,500 years ago that the sea began to approach modern levels and circulation. Once the modern currents were established, plant material could have drifted (or even been carried across the sea) from the southern shores of the Black Sea, where Mediterranean taxa were abundant. This argument has two main premises: (1) oftentimes tree-trunks with incrustations of rocks non-existent in Crimea float ashore; (2) most Mediterranean elements in the pollen diagrams from southwestern Crimea—the only existing pollen records in the Sub-Mediterranean vegetation belt of Crimea's south coast—do not show Mediterranean elements before 7,500 years BP. The first premise is further supported by data published in the same article (Cordova 2007) from a count of floating trash that was washed ashore by a strong storm in the spring of 2000. Labels on the floatable materials linked them mainly to Turkey. Although support for the premise is weak, the evidence shows that sea currents are a possibility for rapid northward migration of thermophilic or subtropical flora.

No substantial paleobotanical evidence exists to support either the postglacial migration of Mediterranean flora into Crimea or their survival in glacial refugia on the south coast. The appearance of Sub-Mediterranean and Mediterranean taxa in southwestern Crimea during the early Holocene is evident in pollen records (Figure 6.1), but the pollen data do not explain whether these taxa arrived in Crimea from lower latitudes or came out of their glacial refugia. Other pollen diagrams encompassing this period are from the piedmont and northern slopes of the mountains (Figure 5.1), where Mediterranean elements are practically absent.

The idea of Mediterranean species surviving in *refugia* during the last glacial phase has also been linked to sea-level changes. Accordingly, during the coldest phase of the last glaciations they were at much lower elevations, perhaps on the shores of the former Euxinian Lake, but as the sea level rose these areas were flooded and drowned, and only a few species may have migrated upward with the rising sea level, as is the case of *Juniperus excelsa* and *Pinus brutia* (Yena, Yena and Yena 2004a). This scenario also explains many of the floristic anomalies, as many species did not survive to become part of the modern vegetation of Crimea or in other cases the creation of a genetic bottleneck, which explains the existence of some neoendemics along Crimea's south coast (Yena, Yena and Yena 2004a).

In addition to the theories recently proposed, explanation of a postglacial migration of Mediterranean taxa into Crimea and/or the survival of Mediterranean species in local *refugia* requires more well-dated paleobotanical data and most definitely genetic studies (Provan and Bennett 2008). In other parts of Europe, the latter have already been applied to identify Mediterranean and other temperate species that survived in glacial *refugia* (Petit et al. 2002).

Endemisms in Crimea

The true character of the endemic flora of Crimea constitutes a scientific issue that has been discussed extensively in the regional botanical and biogeographic literature. This issue has had an impact on the ever-changing number of endemic species in Crimean flora. During the past century alone the number of endemic plant species has been highly variable, with estimates of the number as low as 10 and as high as 300 (Yena 2007). These numbers have recently been revised and the estimate has been narrowed to 106 endemic species (Yena 2012). There are two main reasons for the discrepancies between the numbers. First, the criteria used to assess the true endemics vary according to their being local to regional assessments; clearly, this makes the number of species unique to Crimea considerably lower than the number of species distributed in a relatively broad area. Second, the isolation of Soviet botany for many years did not allow, and reduced the possibility for, comparison of data with other parts of Eurasia.

More recently, geographic standards for local and regional endemics have been established for the broader continental context. These standards include, for example, comparing Crimea with areas of similar size in the Euro-Mediterranean region (Yena 2007). This comparison has enabled the identification of narrow endemisms, which in turn are categorized by their local and regional importance and by their relative age of divergence. The latter comprises the so-called neoendemics and paleoendemics. Accordingly, narrow endemics can be classified into the following categories: local neoendemic and local paleoendemic and regional neoendemic and regional paleoendemic.

Neoendemic species, which were speciated in the recent past, have close relatives in the region. Paleoendemics, which were speciated in the deep past, have few or no relatives in the region. The former are the most abundant in Crimea, which suggests that Crimea's floristic endemisms are a more recent phenomenon, possibly occurring during the late Quaternary, rather than in the Tertiary as previously thought.

The 106 narrow endemics in Crimea are spread into 33 families, of which the most numerous are Asteraceaeae (24), Rosaceae (21) and Fabaceae (11), which in turn are the families with the largest number of species in the Peninsula (Yena 2012). The genera with the highest number of species include *Anthemis* s.l., *Centaurea*, *Minuartia* s.l. and *Crataegus*, with more than five species each (Yena 2012). All but two endemic plant taxa are found in the Greater Crimean Mountain Region (GCMR), which is the area with the largest total number of plant and animal endemisms (Figures 4.8 and 4.9).

Although 106 narrow plant endemic species is a relatively low number compared with other areas of the Euro-Mediterranean region, it is a relatively high number for Ukraine. In fact, the (GCMR) has the highest number of endemisms of any region in Ukraine. In part, the GCMR shares many common traits with the Mediterranean region, one of them being the high number of endemisms (Yena 2012). On the other hand, the high diversity of landscapes in the GCMR, and its geographic isolation from other mountainous areas in the region, are two important factors in the process of floristic endemization.

Crimea: a biodiversity hot spot?

In addition to the concentration of endemic species, the Greater Crimean Mountain Region contains the largest diversity of plants and animal species in the peninsula (Figures 4.8 and 4.9). The estimated number of vascular plant species in the peninsula is 2,536, of which 90 percent lie within the GCMR (Yena 2012). Furthermore the GCMR is the area with the highest biodiversity in the Republic of Ukraine (Biodiversity Support Program 1999; Yena 2007, 2012). But placed in a broader geographic context—that is to say in relation to the Balkans, Caucasus and the broader Mediterranean floristic realm—Crimea's relative high biodiversity looks diminished. This puzzling phenomenon has an interesting explanation, derived from the same biogeographic issues discussed above (i.e., endemisms, land bridges, relicts and anomalies).

The Biodiversity Hot Spots, or areas of high species diversity, mainly occupy two of the world's biomes: the Humid Tropics and the Mediterranean Climates (Myers et al. 2000). On the fringes of these two biomes, other areas have been identified as Biodiversity Hot Spots, among which the Caucasus and the Balkans figure as the most prominent in the vicinity of Crimea. In part this has to do with the fact that the Caucasus and the Balkans share many biogeographic characteristics with the Mediterranean region (Yena 2007). The Ponto-Euxinian floristic province, of which the GCMR is part, also includes a large part of the Caucasus region (Didukh 1992). This situation raises one interesting question for Crimea: if the GCMR's flora has strong affiliations with the Mediterranean, the Balkans, and the Caucasus, why is it not considered a Biodiversity Hot Spot, or at least part of the Caucasus Hot Spot? The answer to this question is complex, but it also alludes to some of the biogeographic issues described such as geographic separation and floristic anomalies.

It is true that many species are shared between the GCMR, the Caucasus, the Balkans and even the greater Mediterranean region. The values calculated by Yakov P. Didukh (1992) state that the Crimean Mountains share as much as 41–60 percent with the Balkans and 61–90 percent with the Caucasus. However, biodiversity refers to the total number of species present in an area, not to the percentage of species shared with other regions. The low number of species compared with the Caucasus and other hot spots is the result mainly on Crimea's separation from the Caucasus (isolation), its size, and its recent geological history (i.e., Pleistocene changes in sea level). Because the GCMR is a geographically isolated entity, the best way to explain these three points is through the Island Biogeography Theory, which is an approach in biogeography applied to the study of species richness in isolated communities.

The Greater Crimean Mountain Region as a "functional island"

The Island Biogeography Theory (IBT) was first presented by ecologists Robert H. MacArthur and Edward O. Wilson in 1967, but since then it has been modified and adapted to several aspects of isolated landscapes, including terrestrial environments (MacDonald 2003). In its general form, the IBT considers that the number of species of an island (in its strict sense) is dependent on two main factors: size and the distance from the mainland. A large island or territory involves, presumably, more diverse landscapes, which means more diverse habitats. Distance from the mainland carries the implication that species turnover in an island can be replaced by new arrivals from the mainland. Thus, in general terms, the supporters of IBT consider that the number of species is directly proportional to an island's size but inversely proportional to its distance from the mainland.

Island Biogeography Theory has its problems when applied to real cases but, overall, it does explain many of the biodiversity traits of islands—and for that matter many isolated ecosystems on land (MacDonald 2003). And despite its practical problems, IBT may help explain biogeographic similarities, and differences, between the GCMR and the Caucasus, the Balkans and the Mediterranean region, although it requires one to assume that the GCMR is an isolated territory surrounded by seas and plains covered by steppe, that is to say, a "functional island." To add the time dimension to this conceptualization, it is necessary to place this functional island in the context of cooling, warming and cycles of sea-level change.

During a glacial stage, when the sea level drops, Crimea's territory has more land connections with the surrounding mainland (Figure 5.4). However, it is important to remember also that low sea levels coincide with the coldest and somewhat drier phases of the Pleistocene, that force many plants to emigrate or survive in *refugia*. Thus, isolation is enhanced as plant *refugia* occupy smaller areas—a process that explains on the one hand, endemization and, on the other, hybridization. The latter is best represented by the case of the Crimean beech (*Fagus sylvatica* and *Fagus orientalis*), discussed in Chapter 4.

A rise in sea level, however, occurs when conditions become slightly warmer and moister. Not only does climatic improvement foster the exit of thermophilous species from their refuges, but the fact that the sea levels are still low (or rising slowly) allows land connections with other regions. This is perhaps what occurred in the Deglaciation period. For Crimea this process in the last ice age should have occurred between 14,000 and roughly 8,000 years ago. But eventually vegetation becomes landlocked in the GCMR as the sea surrounds most of the peninsula, and the northern connection becomes salinized as the water table rises and reaches the surface under conditions of a warmer and drier climate. At this point the GCMR is surrounded and any connections should occur only via sea currents and transport by migratory birds. The functional isolation continues through the interglacial, although in a form different from the one that existed during the last glacial stage.

Into this process of glacial and interglacial isolation enters another important factor: floristic anomalies (Yena 2012). Many species cannot cope with the isolation imposed by cold, dryness or, conversely, by the rapid changes in the sea level. Consequently they do not get re-established in Crimea, which leads to the absence of some species and, in turn, reduces biodiversity. In contrast, in the Caucasus—a region larger and more diverse than Crimea—changes drive the species elsewhere, where habitats are more suitable.

The idea of the GCMR acting as a functional island has important consequences for the formation of endemisms, as some species become isolated from their mother species. Although the concept of a functional island, drawn from the IBT, explains the basis for understanding Crimea's natural biodiversity, endemics, relicts and floristic anomalies, it still requires more studies involving paleobotanical records (i.e., palynological) and genetic research.

Ecosystem restoration targets

Conservation agendas throughout the industrial era have focused on universal issues such as deforestation, extinction of species, destruction of habitats and, more recently, the loss of biodiversity. The issues have always been targeted by action programs, which range from setting aside areas for protection to restoration and/or rehabilitation of degraded ecosystems. The process of ecosystem restoration relies on historical ecology, whose task is to gather information on the ecosystem prior to its destruction. This task involves not only archival information but also paleoecological data (tree-rings, pollen, paleontological data, etc.). In cases where such information does not exist, restoration relies on rules of common sense, and matching flora with soils, and fauna with flora.

In vast areas of North America, Australia, and even some parts of south Saharan Africa, ecosystem restoration programs often target landscapes pre-dating the European colonization. In many cases, those pre-European landscapes were inhabited by hunter gatherers or by groups practising very small-scale agriculture. However, in the case of Europe, and in particular the areas around the Mediterranean, setting restoration targets is difficult because of the long-term effects of agriculture. If the target is to restore landscapes to pre-Neolithic times, then the difficulty is to re-create landscapes that would have been under the influence of climatic conditions different from today's and with flora and fauna that in many cases are not present anymore. If, on the other hand, the target is to restore the landscape to pre-industrial times, then the issue is whether the landscape was being managed sustainably by its local population.

In Crimea the problem of restoration targets presents a duality issue. While the south has seen transformation through time, particularly since the Neolithic period, in a manner very similar to the Mediterranean region, the north has seen radical transformation only recently—perhaps the past 150 years—and in a profound way only in the past 50 years.

If the Crimean steppe landscapes of the plains are to be restored, the target is easy to assess, because enough information on the original landscapes exists from the time previous to the Virgin Lands Campaign (Figure 4.3). This does not mean that all the steppes had never been plowed. A period of optimal moisture in the First Millennium BC allowed many regions to be plowed in vast areas of the Tarkhankut and Kerch peninsulas (Figure 7.1). However, after more than 2,000 years of abandonment, the soils in many of these farmed areas have recovered—if we go by the rate of recovery of a steppe soil in Crimea, which should be in the order of 1,000 years (Lisetskii and Ergina 2010).

Southern Crimea (mountains, piedmont and south coast) seems to be more in the situation of the Mediterranean region in that more permanent settled populations transformed the landscape even before the widespread farming brought about by the Greek colonization. The transformation should go back to Neolithic times, sometime between 6,000 years BP and 5,000 years BP. Even at that time, yailas had been transformed, as shown by the Yaltinskaya Yaila section (Figure 6.2) and the evidence of Neolithic archaeology at high elevations (see discussion in Chapter 6).

But despite the well-known target for restoration in the steppes, restoration to the original vegetation is often not done. One obvious example is the afforestation of the originally tree-less ecosystems in the plains and the yailas. The original vegetation of the plains consisted of a variety of herbaceous communities, none of which included trees (Figure 4.3). Pollen data suggest that the original yaila vegetation was mainly herbaceous; arboreal pollen rarely exceeds 15 percent of the total (Artyushyenko and Mishnyov 1978; Cordova et al. 2011; Chapter 6).

Afforestation of the steppes in the plains is often done by planting tree species not suited for the environment, mainly with native Crimean species that belong to higher levels in the mountains (e.g., *Quercus rubra* and *Pinus pallasiana*). The result is that many of the plantations fail either because of the wind, the dryness and/or the high concentration of salts in the soil. Alternatively, more resistant non-native trees, particularly *Gleditsia triacanthos* and *Robinia pseudoacacia*, are planted with relatively poor success and with the problem of increased competition with the native herbaceous species. Ultimately, however, these non-native trees compete with the native herbaceous species driving them to extinction. The problem also affects the habitats of steppe-adapted fauna.

Similarly, the afforestation of the yailas with species often rare in the area, or with non-native species, has shown failures detrimental to the native ecosystems. Plantations of birch (*Betula pendula*), a native that is extremely rare in the region, have contained a number of dead trees, or trees with trunks bent over by the strong winds. Birch is a relict species of the last glacial stage that today survives in a small area next to the Babugan Yaila, where it is protected by winds. Plantations with local varieties of pine, however, present various degrees of success, but they only create competition with local herbaceous species and increase the material available for wildfires. A similar situation occurs with non-native conifer trees, such as fir and spruce.

The situation in Crimea, as well as in the rest of Ukraine, is that there is no government policy directly addressing the steppes as opposed to the forests (Vasilyuk 2009). Seen from the perspective of restoration ecology, what happens in these cases of steppe afforestation is that re-vegetation is often seen as a substitute for ecosystem restoration, reclamation, or rehabilitation. These terms do not have the same meaning, however. According to Sigurdur Greipsson (2011), restoration means restoring the degraded landscape to its "pristine" or original ecosystem (which, as discussed above, is controversial in areas that have been under transformation since the Neolithic period). Rehabilitation means the restoration of the landscape without the original ecosystem as the main goal, but, rather, one similar to the original where the balance between flora and fauna is re-established. Reclamation involves transforming areas such as former landfills, mines and gravel pits into areas suitable for wildlife. Often, reclamation does involve restoring an ecosystem to a state fairly similar to the original one. Re-vegetation includes simply the establishment of plants—not necessarily native—for commercial reasons, for soil conservation, or aesthetic/visual purposes. In reality, it is the latter that best describes the afforestation of steppes in the plains and the yailas.

The situation with regard to replanting trees sometimes involves many problems, wherever the planting occurs. In Crimea areas are often terraced and trees planted. The terracing, in the first place, makes the landscape look unnatural and does not necessarily slow erosion because herbaceous vegetation is not restored. Pine trees are often used for reforestation but, as known from different studies, pine (unlike oak) does not help curb erosion (Mishnyov 2005).

Similar problems with conservation via reforestation have occurred in several parts of the Mediterranean, where the well-established maquis and phrygana are destroyed and planted with trees. Although both communities are the result of long-term land degradation, particularly pastoralism, they are perhaps the best vegetation cover in the Mediterranean to curb erosion (Hempel 1987). Similar aspects of conservation with native secondary communities could be applied to the south coast of Crimea. In the Karan' Plateau conservation area, on the south coast of the Heraklean Peninsula (discussed in detail in Chapter 11), undisturbed native herbaceous communities are doing the job of holding the soil in place, and in some areas reforestation is taking place as a natural succession.

CHAPTER 11

CONSERVATION ISSUES AND PROTECTED AREAS

A brief history of nature conservation in Crimea

As with the rest of the world, interest in nature conservation in Crimea has existed for some time but it is only in recent decades that it has achieved a high profile. The notion of nature reserves in Crimea goes back to the late nineteenth century, when the term had more of a "game reserve" meaning often linked to the hunting activities of the privileged class. The idea of endangered species appears in the literature at the end of the nineteenth century and in the early twentieth century (e.g., Bumbyeret al. 1913), but nature protection and conservation in its current form began to gain in profile only in the 1920s, gathering momentum in the 1970s when nature conservation, awareness and action developed in tandem with conservation ideas at a national and international level. Today, conservation has become a complex (scientifically and legally) system of nature protection with many controversial issues, particularly in relation to population growth and the conflict with economic growth.

The establishment of protected territories in Crimea began with the creation of an imperial hunting reserve in the Crimean Mountains in 1896 at the location where, in 1870, a hunting cabin had been established near Kosmo-Damianovsky Monastery (Yena, Yena and Yena 2004b). In 1913 a large territory (3,334 ha) inside the area that is now part of the modern Crimean Mountain Nature Reserve was declared an imperial hunting ground (Yena, Yena and Yena 2004b). Although the creation of reserves for hunting was undertaken for the sake of the nobility and the privileged class, over time they effectively evolved into protected areas for animal species that were becoming rare. It was also during this period that ideas about restoring the extinct species began to be put into practice, becoming more common during the first decades of the twentieth century—as demonstrated, for example, by the reintroduction of bison and wild boar (Dulitskiy 2001).

After the February 1917 Revolution, a commission for the management of forests was created, which in 1918 considered extending the existing area of the former imperial reserve in the mountains by adding a further 3,000 ha (Yena, Yena and Yena 2004b). The creation of nature reserves in the modern conservation sense came about in 1923, giving rise to what later became the Crimean Nature Reserve. Other reserves, such as Kara-Dag in 1924, were set up with the aim of studying volcanic structure, but it did not attain the status of nature reserve until 1979 (Yena, Yena and Yena 2004b).

The Great Patriotic War (World War II) had a major impact on the forests in the mountains as fauna were over-hunted by partisans, hungry peasants and Nazi occupiers, and wood was desperately cut for fire (Dulitskiy 2001; Yena, Yena and Yena 2004b). Some of the species reintroduced earlier in the century, such as bison, were killed off (Dulitskiy 2001), whilst in other areas, particularly in the Sevastopol region, constant battles damaged the forests.

After the war, natural monuments such as caves, canyons, and tree groves were set aside for protection. In 1949 the Swan Islands (*Lebyazh'iye Ostrova*) in the Gulf of Karkinitis were declared a reserve. In 1957 the Crimean reserve in the mountains (the one expanded in 1918 and 1923) became known as the Crimean Hunting Reserve, and it is not until 1991 that it was again declared a nature reserve (Kostin and Dulitskiy 1998). By the 1970s, the Mys Martyan, the Yalta Mountain Forest and the Kara-Dag had all been declared nature reserves. Today there are six nature reserves in Crimea and dozens of other protected territories (Figure 11.1). In terms of area, the protected territories cover 5.2 percent of the total territory of the Autonomous Republic of Crimea, and 30 percent of the territory of the Municipality of Sevastopol, which numbers are considerably higher than the figure of 2 percent coverage that applies for the whole of Ukraine (Pozachyenyuk 2009b).

Conservation in the scientific context

The scientific community in Crimea has played an important role in the conservation history of the peninsula, particularly in the last quarter of the twentieth century and the beginning of the twenty-first century. This growing interest in nature conservation has not only seen a number of publications leading to support for more protected areas, but it has also been a means of creating awareness about endangered species and communities.

Interest in conservation has also prompted studies of plant communities generally from an ecosystemic perspective, which in late Soviet times developed into a strong interest in so-called landscape geography

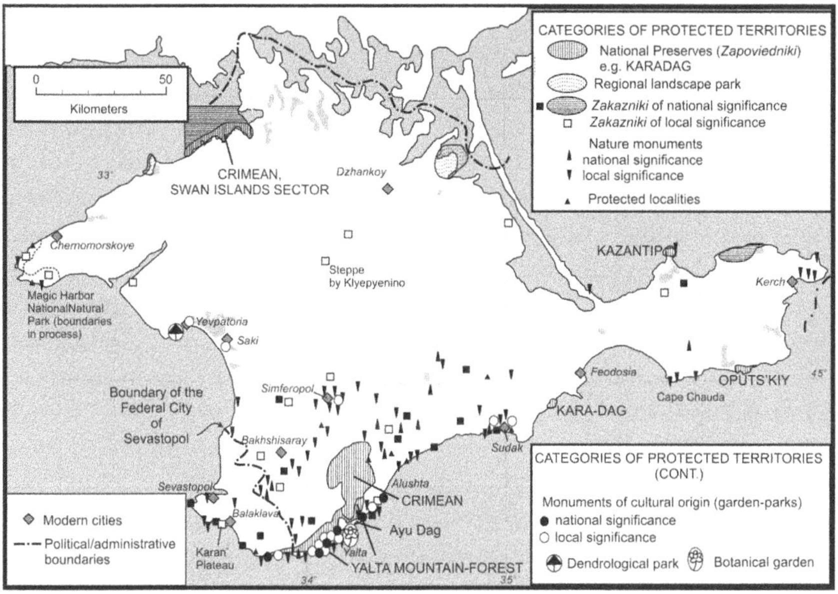

Figure 11.1: Protected territories in Crimea. Source: Karpyenko et al. (2003) with modifications by the author.

(*Landschaftovedeniya*) and geoecology, both of which were important areas of physical geography in research and higher education centers in the former Soviet Union, as well as some Eastern European countries. This approach has been the research focus of many geographers in Tavrida State University (Yena, Yena and Yena 2007). These studies, strongly based in theory, have evolved over the last decade into mapping techniques that use modern mapping technology and Geographic Information Systems (GIS). The concept of landscape has thus evolved, creating concepts more applicable to conservation and leading to the creation of the Crimean Econet, which consists of a series of maps updated with different kinds of environmental and conservation data useful for the determination of fragmented ecosystems and potential corridors between them (Pozachyenyuk 2009b).

Scholars involved in ecological studies of biotic populations and communities began working on a larger cooperative effort which in the 1990s culminated in a series of events that strengthened an awareness of nature conservation issues, particularly threats to biodiversity and endangered species. In 1994, the young independent government of Ukraine ratified the Convention on Biological Diversity, which led to the elaboration of a solid program for nature conservation (Apostolov, Bokov and Dulitskiy 1998). Thus, in 1997 the Biodiversity Support Program with the sponsorship of the World Wildlife Fund, the Nature Conservancy, and USAID organized the

Gurzuf Workshop, where 97 participants met to discuss, and propose solutions to, Crimea's biodiversity and conservation problems. The workshop included a series of exercises that resulted in the identification of, and setting of priorities for, conservation. The scientific results were published in a multi-author monograph in Russian (Apostolov, Bokov and Dulitskiy 1998); an extensive summary of the issues together with proposed solutions were published in a multi-language (English, Russian, Ukrainian and Crimean Tatar) single-volume work (Biodiversity Support Group 1999). A series of scheduled meetings focusing on protected territories followed the Gurzuf meeting in 2001, 2002, 2005, 2007, 2011 and 2013. Each meeting produced a unanimous resolution containing the priorities for conservation and protection directed to government officials.

The post-Gurzuf years have meant not only the positioning of Crimea's biodiversity and natural riches in the national and international spotlight, but also the integration of research and cooperation among local and foreign specialists. Among the new international developments in the area of biological conservation and biogeography has been the inclusion of Crimea in global database collection projects such as the Euro-Med PlantBase and the Atlas Flora Europeaea, as well as the inclusion of certain areas of Crimea into special international programs, such as the registry of the Sivash as a Ramsar Site in the Wetlands International (Yena, Yena and Yena 2007).

Protected territories

Protected territories in the Republic of Ukraine are classified according to a hierarchy based on the purpose and level of protection as well as on activities such as research and tourism. The hierarchy evolved in the context of Soviet law and international directions led by the International Union for the Conservation of Nature (IUCN), the Man and Biosphere Project, and subsequently by the environmental laws of independent Ukraine (Cordova, Rybak and Lehman 2003). The levels comprise different types of protected territory (Figure 11.1). Nearly all types of protected territories exist in Ukraine, with the exception of Reserve Biosphere. Crimea has one national park, six nature reserves, 18 *zakazniki* and 74 natural monuments.

National nature parks

The first—and thus far only—national park in Crimea, the Magic Harbor National Natural Park, was created in 2009. The name "Magic Harbor" (*Charivna Havan'* in Ukrainian) was given directly by Viktor A. Yushchenko, then President of Ukraine. Upon its creation, it was stated that the park would include 6,150 ha already in government ownership, plus a further

4,750 ha from land appropriation, totaling 10,900 ha (Secretariat of the President of Ukraine 2009). The territory of the park comprises a series of existing protected territories at the level of *zakazniki*, natural monuments, and natural objects, including a large area of petrophytic steppe, coastal features such as cliffs, spectacular coastal erosion features such as the Arch of Atlyosh, as well as a number of aquatic environments.

Nature reserves

In Ukraine, nature reserves are territory reserves (*pryrodni zapovidnyky* in Ukrainian; *prirodnyye zapovyedniki* in Russian) set aside for the protection of biotic communities and species as well as for research and monitoring. They occupy the highest level of protection in the hierarchy. The six Crimean nature reserves vary considerably in size, terrain and in their level of complexity (Figure 11.1).

The Crimean Nature Reserve is the largest in Crimea, comprising 44,000 ha and covering various vegetation zones: Mediterranean shrublands and woodlands, southern slope forests, yaila steppes, and parts of northern slope forests. Its size, however, presents some problems, particularly in terms of enforcing protection, for there are no fences around it. The Crimean Nature Reserve was the first protected territory created as a nature reserve and its foundation goes back to the early days of conservation. One section of the reserve—the Lebyazh'iye Ostrova (Swan Islands)—off the northern shore of the Tarkhankut Peninsula, is totally separated from the core reserve. It includes the islands proper and their aquatic surroundings (Figure 11.1).

The Yalta Mountain Forest Reserve, created in 1973, comprises mostly the central part of the southern slope forests and the adjacent Yaltinskaya and Babugan yailas. Its size (14,230 ha) makes it the second largest reserve in Crimea, but as with the Crimean Nature Reserve, it has problems caused by the lack of fencing and the effects of uncontrolled tourism and vandalism. Its proximity to the urban area of Yalta also means that the number of tourists is relatively high.

The Mys Martyan (Cape Martyan) Nature Reserve, also created in 1973, is located just east of the Nikitsky Botanical Gardens. With an area of 240 ha, it is the smallest of all the reserves. It contains the best preserved example of Sub-Mediterranean woodland as well as a number of endangered species, so the reserve has been used for extensive research on Mediterranean and Sub-Mediterranean vegetation. Its smaller area means it is better protected from vandalism and uncontrolled tourism.

The Kara-Dag Nature Reserve, declared in 1979, occupies a volcanic massif on the western part of the south coast between Sudak and Koktebel'

(Figure 11.1). Its size, 2,874.2 ha, makes it the third largest reserve in Crimea. Better protected by its location, the area contains many interesting natural features: a well-preserved volcanic structure of Jurassic age and an interesting array of Mediterranean and semi-arid steppe vegetation.

The Kazantip Nature Reserve and the Oputs'kiy Nature Reserve, were both created in 1998. They are the only two reserves in the Kerch Peninsula and they are youngest in Crimea. The Kazantip reserve has a total area of 450.1 ha, making it the second smallest reserve in Crimea. It comprises a *tombolo* (a rocky promontory linked to the coast by a sand spit) and is an area with interesting geological, flora, land faunal interest. The Opusky reserve occupies a coastal strip on the southeastern corner of the Kerch Peninsula. It comprises 1,592.3 ha of steppe, coastal cliffs and aquatic environments.

Zakazniki

The word *zakaznik* (or *zakaznyk* in Ukrainain) has no direct translation into English, although it may be considered a type of nature reserve established with a specific target for protection. The word comes from the term *zakaz* (meaning "order" in both Russian and Ukrainian), which suggests that the territory has been set aside by a special government order, although in reality this is a somewhat ambiguous idea since all protected territories are set aside this way. In Ukraine the *zakazniki* are meant to protect natural components and complexes, such as communities of species as well as particular species requiring protection (Yena, Yena and Yena 2004b). Unlike the nature reserves, certain recreational activities are allowed. Under Ukrainian law, *zakazniki* are divided into two categories based on their national or local importance. In Crimea there are 17 of national importance and 18 of local importance. Those of local importance are further subdivided by their scientific significance, i.e., floral, faunal (e.g., ichtyological and ornithological), hydrological, geological and landscape complexity. Unlike the level of natural monuments, the *zakazniki* occupy a larger area so as to include territories of more complex landscape composition. Some *zakazniki* territories may even include terrestrial and/or aquatic components.

Examples of *zakazniki* include: Cape Aya and Novy Svyet located on the south coast, where not only several geological features exist but also certain rare and endangered species are protected, as is the case of the Stankiewicz's pine; the Grand Canyon, on the upper reaches of the Bel'bek River, which has both geological and botanical significance, one reason it is referred to as a landscape of significance; and the Karkinitskiy zakaznik, on the northern coast of the Tarkhankut Peninsula, which has a predominantly ornithological value.

Natural monuments and other small protected units

The natural monuments are somewhat similar to the *zakazniki* in terms of protection, allowed activities, and classification by national/local and scientific importance. But unlike the *zakazniki*, they usually encompass smaller territories. They often comprise a feature in the landscape—such as a cliff, a waterfall, a cave or a grove of trees. In Crimea there are 12 of national importance and 62 of local importance. The latter are often subdivided by the aspect they are intended to protect: geological, hydrological, botanical, zoological or landscape, in general. In this context, "monument" is a direct translation of the Russian word *pamyatnik*, which in turn derives from the word *pamiat'* (memory or remembrance). Thus, monument here refers not only to monumental structures but also to something that remains from the past or should be remembered, or more properly is a relic of a past natural landscape.

Two examples of natural monuments are the remains of unplowed steppe near Klyepyenino and Grigoryevka (Figure 11.1), which constitute remains of the original steppe, where not only the community and species are protected, but also where research on natural processes in the steppe is carried out. Cape Chauda, another natural monument on the south coast of the Kerch Peninsula, is predominantly an area of geological significance for the study of Pleistocene marine deposits and terraces.

Other small protected territories include gardens and park monuments of cultural importance, botanical gardens and arboreta. With the exception of the botanical gardens, of which there is only one in Crimea, these units are numerous in the peninsula (Figure 11.1). Unlike the *zakazniki* and natural monuments, these units are intended to protect spaces where plants and animals may exist, but which have been deliberately created by humans—hence their cultural, rather than natural, importance. But like the *zakazniki* and natural monuments, they are also classified according to their importance—national or local. The gardens and park monuments of cultural origin include, for example, several parks along the south coast, some of which were created around mansions and palaces, as is the case with Livadia Park. They include local flora, naturally in place or planted, as well as a number of introduced species. Arboreta (pl. arboretum), which are referred to in Russian as *dendroparki*, are plantations of trees, sometimes containing both non-native and native species. Interestingly, the native tree species may not necessarily be from the part of the peninsula in which they are planted.

Endangered species

Human development and the effects of climate change create pressure on plant and animal species and their habitats. This leads to threats, endangerment and eventually extirpation of species in a particular area; in extreme cases total extinction may result. To address this issue, the International Union for the Conservation of Nature (IUCN) has suggested the creation of Red Lists of endangered species or species requiring protection. Under scientific scrutiny, species are added to or removed from the list.

The Red List, however, is not a legal document that enforces the protection of endangered flora and fauna. Rather, it is a document that sets the scientific basis for legal action (Kostin 1998). Once on the Red List, provisions are made to protect endangered or threatened species through legal procedures, which in some cases result in restrictions on hunting or collection, and in the creation of protected territories in their habitats.

The Red Data Book of Ukraine (the official name of the Ukrainian national Red List) comprises two volumes: one for flora and one for fauna. The total number of plant species in the book is 511; of which 204 grow in Crimea, and, of these, 173 are vascular plants (Vakhrushyeva and Yena 2003). The rest are mosses, lichens, algae and mushrooms. The number of animal species listed in the book is 382, of which 214 live in Crimea. Of these, 21 are mammals, 46 are birds, 7 are reptiles, 19 are fishes and cyclostomes (lampreys and hagfishes), 100 are insects, 4 are mollusks, 11 are crustaceans, 3 are arachnids and myriapods, and 3 are annelids and nematods (Efetov 1999, modified and cited in Yena, Yena and Yena 2004b).

As prescribed by the IUCN, the species in the *Ukrainian Red Data Book* are divided into categories based on the level of threat: Category O includes extinct species; Category E, species under threat of extinction; Category V, species that are vulnerable because they occupy an endangered habitat; and Catergory R, rare species. Using these categories, scientists at the Nikitsky Botanical Gardens have classified the endangered plant species of Crimea. The results by number and as a percentage of the total are as follows (Golubyev 1998):

Category O: 31 species (1.12 percent)
Category E: 41 species (1.44 percent)
Category V: 125 species (4.51 percent)
Category R: 439 species (15.83 percent)

The study mentioned above showed that the total number of vascular plants in the four categories is 636 (22.9 percent of all the species in Crimea)—too high a number compared to the 173 species in Crimea listed in the *Ukrainian Red Data Book*. In view of this problem, local scientists have created a *Crimean Red Data Book* (Kostin 1998), which despite the lack of official recognition has, along with the *Ukrainian Red Data Book*, been influential in achieving recognition of conservation priorities focused on the protection of biodiversity and the necessary steps for action (Biodiversity Support Program 1999).

Protected territories: Prospects and proposals

Although the number of protected territories has increased considerably in the past three decades, their total area comprises only 10 percent of the entire territory of the peninsula (Yena 2007). The protected areas are also unevenly distributed among the major ecosystems (or vegetation regions) of Crimea. The percentage of total area covered by protected areas is highest for the yaila steppes and meadows (almost 100 percent of their territory) and lowest for the true steppe (7.35 percent of its territory) (Cordova, Rybak and Lehman 2003). The percentage of area covered in the other ecosystems is as follows: 96.3 percent for the northern slope forests; 83 percent for the southern slope forests; 67 percent for the Sub-Mediterranean shrublands and woodlands of the south coast; 20 percent for the saline and semiarid steppe, and 11.9 percent for the forest-steppe (Cordova, Rybak and Lehman 2003). It is interesting to note that the saline semiarid steppe, due to its inherent problems for agriculture, has been more successful in attaining protection than the true steppe.

A number of proposals for the creation of new protected territories, or the expansion of the existing ones, have appeared in the past two decades. One of the most publicized is the creation of the Tavrida Natural National Park, by Vladimir, Andrey and Alexey Yena (Yena, Yena and Yena 2004b). The proposed area for this park encompasses a large territory (250,000–300,000 ha) including large portions of the south coast, the northern and southern slopes of the mountains, and the yailas—a vast area that includes a number of existing protected territories (Figure 11.1). Additionally, the proposal includes exclaves of the park in the Kerch and Tarkhankut peninsulas. The areas proposed for the park include territories that have the richest endemisms and biodiversity areas in the peninsula (Figures 4.8 and 4.9). The project also combines many aspects of protected areas (especially the nature reserves) and possibilities for controlled tourism. Inspired by some of the natural trails, such as the Appalachian Trail in the United States, the project encompasses a Crimean Trail (Yena, Yena and Yena 2004b).

Other proposals for the creation of natural national parks exist for other areas, including the Sivash (discussed in the next section). Similar proposals for parks exist in other parts of Crimea, as is the case, for example, of the Heraklean National Natural Park, which is meant to include many existing territories in the Sevastopol municipality (Litvinyenko 2010, Sevnews.info). Efforts among local, national and international advocates continue to expand, one way or another, the coverage of protected areas. Although the creation of the Magic Bay National Natural Park was a great event for the protection of Crimean nature and environment, it ignored many of the other existing proposals to create natural parks in Crimea.

Protecting the Sivash

Although a young lagoonal system on a geological time scale (Chapter 6), the Sivash and its surrounding features (e.g., the Arabatskaya Strelka) have developed a series of habitats for plants and animals, both terrestrial and aquatic. But despite being in an area of relatively low population density in the peninsula, the habitats of the Sivash are among the most endangered in Crimea. The western basin of the Sivash has been affected most by human activity, having been transformed by a series of evaporation dams linked to the extraction of salt minerals. The other basins are less modified, but they are exposed to the effects of agriculture and irrigation.

Among the most important threats faced by the Sivash from agriculture and irrigation is pollution. In large part this stems from agricultural development in the surrounding areas, where waters contaminated with fertilizers enter the lagoonal system via rivers, canals, and groundwater fed by irrigation and canal leaking. This results in the frequent poisoning of birds by the chemicals used in agricultural activities (Siokhin and Kostyushin 2000). Birds are not the only biotic group threatened by human activity around the Sivash: the area also has a number of other species listed in the *Ukrainian Red Data Book*: 5 insect species, 1 fish species, 2 reptile species, 22 bird species, 2 mammal species and 7 plant species (Gorodyetsky and Karpyenko 2000).

Some achievements have been realized in terms of conservation in the Sivash. First, a number of protected areas, though small, have been established in the area (Figure 11.1). Second, the Sivash is listed as one of the wetlands in the Ramsar Convention on Wetlands, which gives it international visibility and promotes research, protection and education. Its international significance lies in the fact that the Sivash ecosystem is an essential part of the network of paths used by migratory birds traveling between Eurasia, the Mediterranean and Africa, as well as a wintering ground, since the highly saline water of the Sivash hinders ice formation (Siokhin and Kostyushin 2000). In winter about 500,000 birds of 25 species have been recorded in the area (Gorodyetsky and Karpyenko 2000).

As mentioned above, the Sivash Natural National Park is still at the proposal stage (Karpyenko and Lychak 2000). Azovo-Sivahsky National Park in the Kherson Oblast', on the other hand, already exists. It only includes, however, the northern shore of the central basin of the Sivash, and the creation of a park on the Crimean side would expand the protected area into the southern parts of the lagoonal system. One potential difficulty for the creation of a park on the Crimean side of the Sivash is that very few protected territories exist there, in contrast to the proposal for the Tavrida Natural National Park, or even the existing Magic Harbor Natural National Park in the Tarkhankut Peninsula.

The scarcity of existing protected areas in the Sivash area is an issue that relates to a further problem: conflict with economic growth, particularly mining and agriculture in the region. Accordingly the creation of a park would need to include economic benefits to offset the losses incurred in these other activities. Perhaps other recreational aspects—such as health resorts which make use of the healing muds, sailing, kayaking and canoeing, among other activities—should be considered (Karpyenko and Lychak 2000). Bird watching has already become an important attraction in the area, and can be considered one activity for tourist development.

Protecting the yaila ecosystem

Among the human threats faced by the yaila ecosystem are the direct and indirect effects of uncontrolled tourism, frequent fires, and inappropriate restoration by planting trees. The importance of protecting these highland ecosystems lies beyond the mere protection of flora and fauna: its protection also concerns the hydrological system of the mountains. Because of the karstic nature of the terrain, the yailas absorb large amounts of atmospheric water, for which its vegetation cover is essential. This water feeds the aquifers of the mountains, which in turn feed springs and river systems to meet the needs of the human population (Olifyerov 2011).

Planting trees in the yailas is not an adequate measure for protecting a landscape that apparently is not suited to forests. A number of tree species that have been planted do not develop properly because of the strong winds and the cold. Moreover, palynological data suggest that the landscapes of the yailas have been naturally treeless (Artyushyenko and Mishnyov 1978; Cordova et al. 2011; see also discussion in Chapter 6). Therefore, the key to protecting the yaila ecosystem lies in the protection of its herbaceous communities. Although grazing in the western yailas has been prohibited for almost a century, in some of the eastern yailas, which are not under the protection of a nature reserve, grazing still occurs to some extent even today.

Ideas for the reforestation of the yailas, based on the assumption that they were once forested, date back to the beginning of the twentieth century, with particular impetus between 1957 and 1987 (Olifyerov 2011). In part, the idea of planting trees in the yailas has strong support from the hydrological point of view, particularly because the aquifers are recharged in the karstic plateaus. And although Avgust N. Olifyerov praises the practice, given its results in increasing the balance of the aquifers, springs and rivers, he considers that more studies in the field of karst hydrology are needed to understand better the true nature of the aquifers and the best ways to recharge them (Olifyerov 2011: 184).

The proposal for afforesting (or supposedly reforesting) the yailas has its basis in the grassy plateaus having traditionally been used as summer pastoral grounds, implying that the pastures were created by human action, and therefore they have to be restored to their alleged original status. But aside from the contradictions that exist between the hydrological rationale for afforestation and the facts shown by paleoecological data, the problems of tree planting in the yailas can be summarized as follows:

1. The practice means terracing and the alteration of the soils, which in turn may have consequences in terms of carbon balance, and removal of material, which may cause erosion.
2. Terracing is implemented in some areas, also creating problems for native species, as well as for the aesthetic natural value of the landscape.
3. Many of the planted species do not thrive well. Many die from the cold and the wind, and those that survive are of short stature, with bent stems due to the strong winds.
4. To overcome the problem of (3) many non-native species have been introduced, including spruce and fir, for which there is no record in Crimea, even for the last glacial stages (see Chapter 5).
5. Trees certainly compete with the native herbaceous vegetation of the yailas, thus reducing soil protection, which in many cases is ineffective because pine—the most common plantation tree—does not protect soil effectively against erosion (Mishnyov 2005).
6. An increase in trees increases fuel for wildfires, which can spread rapidly under the constant windy conditions in the yaila.

In summary, tree planting in the yailas seems not to be a solution to erosion and not the best solution, ecologically, to increasing water recharge. Soil erosion in the yailas is not necessarily caused by deforestation, since it seems to have been common even at the time the first herders began using their pastures for livestock. Uncontrolled tourism, on the other hand, is a current problem, which not only has an impact and causes stress on vegetation, but also intensifies the frequency of fires and promotes soil erosion. Driving off-road, a common happening with visitors, causes compaction of the soil and therefore a reduction in infiltration. Given these problem, perhaps protecting the herbaceous cover of the yailas could be effective for both aquifer recharge and for the protection of the native flora and fauna.

The Karan' Zakaznik: A recent effort in nature protection

The creation of one of the most recently protected areas in Crimea—the Karan' Zakaznik—is connected to an interesting historical and ecological process that is worth mentioning. The Karan' is located on the southern heights of the Heraklean Peninsula, between Balaklava Bay and Cape Fiolyent. The territory of the zakaznik comprises: the Karan' plateau, made of relatively horizontal layers of the Miocene calcareous marine rocks of the Sarmatian Series; a hilly area on the east; and a cliff that exposes Miocene calcareous rock strata, Jurassic marmorized limestone, and volcanic rocks. Because of its strategic position, the area was included in a military base that occupied large parts of the territory of Cape Fiolyent and the former Gregorian monastery.

The soils of the Karan' are mainly stony calcareous (rendzina soils), and calcareous cinnamonic. Although remains of Greek agricultural installations are found in the area, and some cinnamonic soils present traces of ancient farming (Lisetskii et al. 2012), the thin soils seem to have been largely used for pastoral activities. The hilly areas, the valleys and the cliffs, support woodlands of *Juniperus excelsa*, *Pistacia mutica* and *Carpinus orientalis*. The plateau is dominated by herbaceous communities of *Stipa lithofila* and *S. braunerii*, *Asphodeline lutea* and *A. taurica* (Pozachyenyuk 2012). Besides being a good example of communities with Mediterranean and Sub-Mediterranean flora, these communities are characterized by a high diversity of species and a large number of rare and endangered species, several of which are listed in the *Ukrainian Red Data Book* (Bondaryeva 2010; Pozachyenyuk 2012).

For several decades, the Karan' plateau was leased to the adjacent military base of the Black Sea Fleet. Unlike other areas in the base, it was rarely utilized for training since its purpose was mainly strategic. Its inclusion in the base gave it protection from agriculture and pastoralism, from tourism and practically any other activity detrimental to vegetation. Thus the area became a *de facto* reserve.

In 2006, the lease to the military base area expired and the Karan' plateau reverted to the city of Sevastopol. Shortly after the lease came to an end, a number of companies put forward a plan to create an upscale resort and a golf course, but thanks to pressure from scientists and local environmental activists, the Municipality of Sevastopol finally declared that the area should be set aside for conservation. Thus, documentation for the creation of a zakaznik, including scientific data, was brought forth. It was proposed that the Karan' area should be designated a zakaznik of local importance with botanical and landscape significance. But despite this designation, attempts to continue with the plans for a golf course have impeded the establishment of the zakaznik (Litvinyenko 2010, Sevnews.info). Besides the conservation area of the Karan' plateau (including the cliff and aquatic environments) other areas of forest and areas surrounding the Balaklava quarry have been added, totaling an area of 1,071.28 ha, of which 124.8 ha constitute an aquatic component (Slava Syevastopolya 2006).

The scientific community took the main steps by outlining the botanical and ecological importance of the area, including the number of endangered habitats and species. Scientists at the Institute of the Southern Seas in Sevastopol have pointed to about 50 species listed in the *Ukrainian Red Data Book*, some of which are also listed in the *European Red Data Book* (Bondaryeva 2010, in pryroda.in.ua). Scientists and the community have worked together in this effort, which is an excellent example of what can be done and offers optimism and encouragement for expanding protected areas in Crimea. At the same time the recent events in the Karan' provides an object lesson of the difficulties to be faced when setting aside territories for protection in areas with strong economic and political interests.

Save the Crimean steppes

The number of protected territories, and the area they cover in the steppe, is very small and scattered (Figure 11.1). By contrast, remains of non-protected, non-plowed steppes cover vast areas of the plains. They include: the steppes of the Tarkhankut Peninsula (1,172 km^2); the Sasyk-Kyzil-Yar steppes (75 km^2); the Skvortsovskiy steppe (110 km^2); the steppe fragments of the Krasnogvardyetsk district (395 km^2); the steppes around the mouth of the Salgir River (110 km^2); the steppes along the Arabatskaya Strelka (150 km^2); and the unplowed steppes of the Kerch Peninsula (1,940 km^2) (Parnikoza and Vasilyuk 2010, web resource). The total area is 3,934 km^2 of fragmented and unprotected steppes.

As discussed in Chapter 9, the main threats to the steppes are area reduction and loss of biodiversity, pollution of water and soil, invasive species from cultivated fields, and afforestation. Afforestation policy has recently been criticized by scientists and activists, as the government adheres strongly to the policy of planting trees in the steppes. This policy in part is a rapid means of meeting the minimum European requirement for conservation: to the bureaucratic mind this can be achieved by transforming large areas of steppe into forest (Vasilyuk 2009). The measure violates the conservation laws of

Ukraine, the *Ukrainian Red Data Book* and the Bern Convention of Biological Diversity, because it threatens the many endangered species living in the steppe ecosystems (Vasilyuk 2009).

The low number of protected areas in the steppes, and the careless management of existing protected territories, as well as the careless afforestation, has prompted the creation of a non-government organization in Ukraine, *Zberezhemo Ukrayins'ki Stepy!* (Save the Ukrainian Steppes!). The organization is composed both of scientists and activists, a collaboration that has made clear the benefits of restoring and rehabilitating the steppes based on historical and paleoecological accounts (Zberezhemo Ukrayins'ki Stepy! 2010, web resource). The effort also includes a broader regional movement encompassing other countries of the Eurasian steppe. Many of the publications on the subject are to be found in the Russian-language publication *Styepnoy Byullyetyen'* (Steppe Bulletin), published by the Siberian Ecological Center. The Save the Ukrainian Steppes organization collaborates with its Russian counterpart, *Sokhranyeniye Stepyey Rossii* (Conservation of the Russian Steppe). But the movement is young and new developments will surely come—probably pressure on government to change policies for landscape rehabilitation. As recently as the summer of 2013, activism and legal procedures stopped large-scale afforestation projects in Crimea (Steppe Bulletin Editorial 2013).

The role of nature conservation

Praised by travelers and scientists, as well as writers and artists, Crimea's natural heritage is unique and worth preserving for generations to come. Crimea's scientists have strived to identify, value and protect the components of the peninsula's natural heritage, which along with its cultural heritage (archaeology, historical buildings and other cultural features in the landscape) constitute one of the most valuable resources for Crimea.

The peninsula is relatively poor in precious minerals and fuels, but is rich in renewable resources such as an attractive climate, visually attractive landscapes and a series of unique natural and cultural features of scientific and aesthetic significance. It is for this reason that the natural and cultural features of Crimea should be protected and not sacrificed to rapid development. It is true that after a turbulent political and economic past, the peninsula is seeing some of its better days, but these days can be perpetuated by protecting Crimea's main assets: its nature and culture.

One of the darkest periods in Crimea's conservation history, which was also the worst for the whole Soviet Union, was the Khrushchev era. Not only did the Virgin Land Campaigns destroy large tracts of a variety of ecosystems, but the premier himself was also highly resistant to conservation, often seeing reserves as economically worthless (Johnston 2012). Such a view is not unique to socialism or to capitalism: it is one that prevails in many parts of the globe. An exception to this, however, has been those areas where charismatic megafauna, particularly in Africa, have made conservation a profitable enterprise, despite many setbacks and controversies.

Assessing, monitoring and protecting endangered species, habitats and ecosystems, and establishing a network for studying the effects of climate change, is one of the priorities for scientists. Local institutions of research and teaching are a resource in terms of ideas and personnel for the study and protection of nature. And the most recent publications, which unfortunately are still not making it into the international arena, show the variety of disciplines—geography, botany, geology and other environmentally focused subjects—available for tackling the main environmental issues and for monitoring nature. As stated in the preface, the purpose in writing this book is to make Crimea's natural and cultural wealth visible to western readers and to the international scientific community. Perhaps colleagues from other countries will become interested in collaborating with local scientists in their endeavor to better understand and protect this fascinating land that is the Crimean Peninsula.

CHAPTER 12

THE MEDITERRANEANIZATION OF CRIMEA

Crimea between the Greater Mediterranean and the Eurasian steppes

Throughout the study of the nature and cultural history of Crimea, it is hard to miss the connection with the Mediterranean region. But in reality, the Mediterranean connection is strong only in the south of the peninsula, mainly the south coast, the mountains and, to some degree, the western half of the piedmont and parts of the west coast. The rest of the peninsula, namely the plains, has much more in common with the Eurasian steppes than with the Mediterranean region. But even if one focuses on that part of the peninsula that has stronger ties with the Mediterranean region, it is hard to evaluate whether or not southern Crimea should be part of the Mediterranean realm. This difficulty arises from the number of criteria used to define the Mediterranean as a geographic region.

If the limits of Köppen's "Cs" climatic type (i.e., Mediterranean climate) are used to define the Mediterranean realm, the Black Sea region is excluded, except for the area around the Bosporus (see Figure 1.6). However, if biogeographic criteria are used, the extent of the Mediterranean realm reaches beyond the Cs climatic limits, as significant numbers of Mediterranean elements are found in relative abundance in many regions of the Black Sea, particularly southern Crimea and the coast of the western Caucasus. It is the biogeographic criterion that has led to the suggestion that southern Crimea is an exclave of the Mediterranean biome (Vul'f 1944; Gratsianskiy 1971; Walter 1974; Rubtsov 1978). This idea of it being a Mediterranean exclave has implications beyond basic biogeography, as the idea has permeated aspects of economic history and the transformation of the Crimean landscape.

As well as the physical features that one associates with the Mediterranean realm, there are the cultural aspects and these extend beyond the climatic and biogeographic boundaries. For example, the extent of grape cultivation has extended to latitudes way beyond the borders of most Mediterranean floristic elements. And the links of Crimea with the Mediterranean through trade, colonization and imperial subjugation, place the peninsula the cultural sphere of the Mediterranean realm.

One quick way to evaluate the links with the Mediterranean is to apply the criteria in Fernand Braudel's classic treatise of the Mediterranean region: *The Mediterranean World in the Age of Philip II* (published first in 1966 and published in English in 1971). First, Braudel uses the term "Greater Mediterranean" region to include large parts of the northern Sahara, the deserts of the Levant, and the area around the Black Sea. Second, he uses two sets of criteria to define the region: the Mediterranean as a physical unit, and the Mediterranean as a human unit.

Should southern Crimea be considered part of the Greater Mediterranean? The following brief analysis uses Braudel's criteria based on a checklist of physical and human traits of the Mediterranean realm. The area considered here is bounded by an imaginary line linking Sevastopol–Alushta–Bakhshisaray–Sevastopol, and the criteria do not include flora, which according to Yakov Didukh (1992) is between 50 percent and 60 percent of the floristic elements of this region.

Physical Unity Criteria

1. Part of the Alpine System: Yes
2. Mean January temperature higher than 6°C at sea level: No*
3. Summer drought longer than three months: No
4. Most precipitation (70 percent) between October and April: Yes
5. Climate influenced strongly by cyclonic waves originating in the Atlantic and passing through the Mediterranean Sea: Yes
6. Thin, stony, carbonated soils: Yes
7. Soil rubification (soil reddening): Yes

* The highest mean January temperature (2°C) is recorded in Yalta.

Human Unity Criteria

1. Traditional pastoralism of the husbandry type: Yes
2. Traditional olive cultivation: No
3. Traditional wine cultivation: Yes**
4. Agricultural terracing: Yes
5. Influence of Phoenician and/or Greek trade: Yes
6. Part of at least one of the following empires: Roman, Byzantine, Ottoman***

7. Twentieth-century decline of traditional rural management: No****
8. Tourism linked to beaches and warm climate: Yes

** Not as widespread but common in areas with high population density in the Middle Ages

*** Crimea was part of all three empires

*** Under Russian and Soviet rule, the rural economy went in a different direction.

The total number of "yes" answers is 5 out of 7 for the natural criteria and 6 out of 8 for the human criteria. In a conservative way, this means that Crimea is not part of the Mediterranean realm. Geographically, it is not; it is in the Black Sea Basin. But the question is: Do all the regions within the Mediterranean region fulfill all 15 criteria above?

Less conservatively, on the other hand, it seems that Crimea shares many characteristics with the Mediterranean region, making it perhaps part of Braudel's Greater Mediterranean Region, whose boundaries—as is the case with all geographic regions—are diffuse with nuances and transitions. Crimea thus seems to be on a transitional boundary.

The Mediterraneanization of Crimea

The process by which Mediterranean life-styles and economies (particularly tourism) in areas outside the Mediterranean region are emulated is referred to as Mediterraneanization (Morris 2006). Often the process is referred to as Mediterraneanization, particularly in the natural science literature (e.g., Cordova 2007; Oğuz and Öztürk 2011). But from the point of view of biogeography and environmental history, the term Mediterraneanization here is used to refer to (1) the physical or natural processes by which Mediterranean plant taxa have become established in Crimea and formed communities similar to those of the Mediterranean region; and (2) the cultural processes that have created a Mediterranean-like landscape economy, not only the deliberate introduction of plants (including cultivars) but also the creation of land management methods and a rural economy similar to that of the Mediterranean and, overall, an economy linked to Mediterranean-like landscapes, including tourism.

The process of physical Mediterraneanization in Crimea began as Mediterranean flora became established in the peninsula during the Holocene, as conditions warmed up and the sea level rose (Figure 6.1 and discussion in Chapter 6). Cultural Mediterraneanization began with the Neolithization of Crimea, and took a faster pace as Greek settlers introduced agricultural practices and crops typical of the Mediterranean region in the First Millennium BC, a process that was perpetuated by other groups (Byzantines, Genoese and Ottoman Turks). The cultural process culminated in recent centuries with the vast production of wine and other Mediterranean products and with the creation of a Mediterranean ambience—referred to as the Russian Riviera, and later the Soviet Riviera, as described in Chapter 8. Ultimately, Mediterraneanization, both physical and cultural, has played a tremendous role in shaping the modern economy as well as influencing aspects of the conservation and management of natural and cultural resources, which is one reason it is an important process in the environmental history of Crimea.

A chronological framework for the Mediterraneanization of Crimea

The best way to understand the chronological evolution of the physical and cultural processes of Mediterraneanization is to keep in mind the factors that contributed to them in space and time. According to Ludwig Hempel (1987) Mediterraneanization began in southern Europe when modern Mediterranean vegetation became established after the last glacial phase. In other words, this process started when the typical Mediterranean flora began to dominate in pollen frequencies sometime around 11,900 years BP (Vogiatzakis 2012). In Crimea, the process of Mediterraneanization of the vegetation occurred later. Although most Mediterranean and Sub-Mediterranean elements in the pollen diagrams of southwestern Crimea begin to appear between 12,000 years BP and 10,000 years BP, they do not become prominent (but not dominant) until around 7,500 years BP (Figure 6.1). The process of cultural Mediterraneanization in Crimea lagged behind its counterpart in the Mediterranean lands, since it was necessary first to build the physical scene and establish the contacts through diffusion and migration. To better delineate the process of Mediterraneanization in Crimea we can identify seven phases.

Phase 1, or the pre-Mediterraneanization phase, comprises the rearrangement of vegetation and the slow expansion of Mediterranean floristic elements from their glacial *refugia* or directly from southern latitudes.

Phase 2 is the beginning of the physical Mediterraneanization, which involves a global increase in temperatures, changes in the sea level, and the conditions for Mediterranean floristic elements to become fully established.

Phase 3 involves the beginning of landscape transformation through pastoralism and later farming, developments which originated in the eastern Mediterranean. These developments led to the transformation of the vegetation into typical secondary communities such as the shiblyak and phrygana that we see today in the south and southwestern parts of the peninsula.

Phase 4 initiated the direct-contact trade and settlement of immigrants from the Mediterranean. This phase begins with the ancient Greek colonization (see

Chapter 7) and constitutes the first step in the full consolidation of the cultural Mediterraneanization of the peninsula.

Phase 5 comprises the continuity of the systems established in the previous phase and the time when Crimea was linked to Mediterranean-based empires: Roman, Byzantine, Genoese and Ottoman.

Phase 6 comprises Mediterraneaization on a large scale and with commercial or institutional purposes, denoted by the time when Mediterranean plants were introduced and viticulture was fully developed. It is the time also when the Mediterranean idea of tourism appears.

Phase 7, which roughly coincides with the post-Soviet era, encompasses the process of Mediterraneanization fueled by globalization. Interestingly, the cultural aspects linked to economic development are threatening the physical part of the process by threatening the natural communities and biota linked to the Mediterranean.

Physical Mediterraneanization

Several geographical and botanical studies have pointed to the existence of a Mediterranean exclave on the south coast of Crimea (Gratsianskiy 1971; Yena 1978; Douguedroit and Zimina 1987; Vodop'yanova 1986; Didukh 1992; Vyed' 2000; Cordova 2007). Often this exclave is referred to as "Sub-Mediterranean," since it differs in many biogeographical and climatic aspects from the core Mediterranean region (Yena 1978). Nonetheless, the climate of Crimea's south coast differs from the Mediterranean in that winter temperatures are lower and summers are not completely dry (Douguedroit and Zimina 1987). However, the same study found similarities with the southern coast of France, located on the northernmost margin of the Mediterranean climate, where winter conditions are cooler and summers are not as dry as in other areas of the Mediterranean. Interestingly, the two regions are located at about the same latitude as the south coast of Crimea (Figure 1.6).

As discussed in Chapter 3, the climatic conditions that keep Crimea's south coast warmer result from the protection provided by the mountains blocking the rolling cold polar masses of air and the influence of a relatively warm sea. These climatic conditions allow thermophilous taxa of Mediterranean affiliation to thrive (Figures 10.1, 10.2 and 10.3). Additionally, certain plants communities on the south coast resemble those of the Mediterranean. The Crimean shiblyak and phrygana have often been equated with the Mediterranean maquis and phrygana respectively, although differences exist particularly in the absence of edificatory species (Yena 1978). The name shiblyak, which is traced to the Dalmatian coast in the Mediterranean, is used to refer to shrubland communities resulting from the degradation of the Mediterranean forests (Rikli 1943; Didukh 1992).

The maquis, unlike the shiblyak, is comprised chiefly of evergreen shrub communities (Rikli 1943). Evergreen elements are present in the shiblyak but they do not constitute the majority of species (Golubyev 1997). However, both the maquis and shiblyak are considered secondary communities resulting from the degradation of forests (Rikli 1943; Didukh 1992; Pozachyenyuk and Grishankov 1998).

The Mediterranean phrygana (known in some areas as *garrigue*) is another secondary community. Although it presents an enormous variety of forms (Rikli 1943) the phrygana communities are dominated by unpalatable species, which vary from spiny types such as *Sarcopoterium spinosum*, a plant that is rare in Crimea, to a variety of other scrub including many species of the Lamiaceae (mint family), which are abundant and varied in Crimea. The Crimean phrygana occurs mainly on calcareous ground, for example the western part of the south coast (Figure 4.6), and mainly at higher elevations in the form of the petrophytic herbaceous communities of the yailas (Didukh 1992; Golubyev 1996a).

There has long been debate among biogeographers, botanists and conservation specialists regarding the original vegetation of the areas now covered with maquis and phrygana (Rikli 1943; Vogiatzakis 2012). It has always been assumed that this was forest, although there may have been different varieties in different stages of ecological maturity since Neolithization began soon after the postglacial development of forests in southern Europe (Hempel 1987; Vogiatzakis 2012). Furthermore, it seems that after the long-term effects of agriculture, and particularly pastoralism, there is now a well-established and stable community but with a high degree of complexity and biodiversity (see, for example, Tomaselli 1977; Naveh 1982).

As for the Crimean shiblyak and phrygana, the situation is similar to their counterparts in the Mediterranean, as they hold a relatively stable structure and composition, as research in the KaraDag and Mys Martyan shows (Shelyag-Sosonko, Osychnyuk and Andriyenko 1982). However, the variety of shiblyak and phrygana types implies that there are different successional stages, some of which can take so long that from a human perspective they seem relatively stable. While some successional paths evolve into different types of shiblyak, others do evolve into forests (Pozachyenyuk and Grishankov 1998).

Pollen records seem to support the idea of a shiblyak caused by Neolithic activity, possibly pastoralism, as typical shiblyak species developed from around 6,000 years BP to 5,000 years BP, replacing forest that had developed between 7,500 years BP and 6,000 years BP (Cordova and Lehman 2003 and 2002). However, pollen taxa of phrygana elements appear as early as 9,000 years BP. It is possible that this early Holocene phryganoid community was a form of a petrophytic steppe responding perhaps to dry and cold conditions, apparently the Younger Dryas, the setbacks of the 8.2 ka and other events.

In addition to vegetation change, soil development also supports the process of physical Mediterraneanization throughout the Holocene. The relatively warm dry summers and wet winters are reflected also in the development of the typical cinnamonic soils, which have been equated with the Mediterranean red soils (Gerasimov 1954). The soils are, in fact, associated with those areas of the peninsula that have a Sub-Mediterranean climate (Dragan 2000) and with the distribution of Mediterranean plant species, particularly with *Pistacia mutica* (Cordova 2007). These areas include the south coast, the Heraklean Peninsula, and the western part of the piedmont (Figure 3.12).

Based on pollen and paleopedological data from southwestern Crimea, the Sub-Mediterranean vegetation and cinnamonic soils of southern and southwestern Crimea began to develop between 8,000 years BP and 7,000 years BP (Cordova and Lehman 2005; Cordova 2007). Between 7,000 years BP and 5,000 years BP, Mediterranean taxa became established, a process that may be related to (1) the establishment of the modern configuration and levels of the Black Sea, and (2) the relatively high temperatures of the hypsithermal period (Mid-Holocene warming) (Cordova 2007). Modifications to its communities and soils by humans developed into the modern Sub-Mediterranean landscapes (Cordova and Lehman 2003, 2005; Lisetskii et al. 2012).

Cultural Mediterraneanization

The process of cultural Mediterraneanization in Crimea evolved with the development of agriculture. The Neolithic revolution originated in the Fertile Crescent, an area that includes the Mediterranean Levant, and from there it spread west and north, reaching Crimea in the Fourth Millennium BC. This first diffusion of agriculture brought some traits already developed in the Mediterranean region, particularly those traits linked to cereal cultivation, orchards and pastoralism, which were widely practised by the Bronze Age people of Crimea (Chapter 6). Nevertheless, the more solid insertion of Mediterranean agricultural elements into the cultural landscapes of Crimea occurred during the Greek colonization (eighth to first centuries BC), particularly through the introduction of grape cultivation and other crops typical of Mediterranean rural systems. As discussed in Chapter 7, subsequent cultures in the region continued with the Greek tradition of wine production, as well as the development of other crops such as figs, walnuts, pulses and cereals, complemented by livestock grazing, particularly goats and sheep. Olive cultivation—the most typical cultural feature of the rural Mediterranean economy—was never introduced on a large scale to Crimea because of the tree's sensitivity to cold winters.

One of the first systematic studies of the acclimatization of olives to Crimea was carried out in the early twentieth century by Yevgenny Wulff and colleagues on the grounds of the Nikitsky Botanical Gardens (Vul'f, Kalayda and Plotnitskiy 1916). The studies looked not only at the old trees on the south coast—for example, the 500-year-old olive tree inside the gardens—but also at a series of historical references and experiments. The idea that olive cultivation was practised by Greek colonists is not supported by the pollen records of the Heraklean Peninsula and Chyornaya Valley (Cordova and Lehman 2003), but it is possible that the tree may have been introduced to the south coast around Yalta, where unfortunately there are no pollen records.

The poor resistance of olive trees to low winter temperatures means that the crop could have been developed only along the warmest parts of the south coast (Vul'f, Kalayda and Plotnitskiy 1916). This area excludes the Heraklean Peninsula, where a well-studied Greek settlement (the chora of Chersonesos) existed. However, in the early 1800s, Peter S. Pallas (cited by Vul'f, Kalayda and Plotnitskiy 1916) reported old escaped trees in the Mangup plateau, where olive cultivation might have been possible during the Middle Ages. No other records support large-scale olive cultivation in this region during late Antiquity and the Middle Ages.

Despite its adaptability to the south coast and the positive results of several experiments with Italian and Spanish varieties, the olive industry in the form of oil or pickled olives never flourished on Crimea's south coast in modern times. Unlike the success of the more adaptable grape cultivation, olive oil failed to compete with imported oil (Vul'f, Kalayda and Plotnitskiy 1916). Furthermore, there were other aspects of agroclimatology and management that also put grape cultivation at an advantage over olive trees and other Mediterranean crops. First, grapevines are deciduous plants—an advantage for harsh winters. Second, the crop is easily adapted to the local climatic conditions of Crimea through hybridization with cold-resistant wild varieties—in fact, not much different from the way grapes were introduced in the burgeoning wine lands of California. Beyond these fundamental aspects, grape cultivation has been managed successfully in areas of the plains, where low winter temperatures and/or seasonal soil moisture availability have always been a problem for grapes. In fact, to this day, most of the grapevines are covered during the winter in most northern and central parts of the peninsula. In these same areas, where soil moisture is not enough, particularly at the critical time of bud sprouting and blooming, irrigation is implemented (Fursa 1977).

The annexation of Crimea by the Russian Empire opened a new chapter in the history of cultural Mediterraneanization, as the idea of introducing Mediterranean plants into Crimea began to appear (e.g., Peter S. Pallas;

Christian Steven). This idea was put into practice through the creation of the Nikitsky Botanical Gardens, which initiated experiments to acclimatize crops and ornamentals from the Mediterranean region as well as other subtropical regions (Shamko 1963). In particular, the introduction of Mediterranean and other subtropical ornamentals, some of which became a success, resulted in the widespread distribution of adventive flora, some of which have even made it to the wild and have become naturalized. Mediterranean trees, such as the Italian pine (*Pinus pinea*), the Portuguese oak (*Quercus faginea*) and the now ubiquitous cypress (*Cupressus sempervirens*), are very common in gardens, parks and alleys, and in many areas they have replaced the native trees.

As the acclimatization of commercial and ornamental plants gained momentum, the idea of Mediterraneanization in the touristic sense emerged. Thus, by the end of the nineteenth century, the Russian Riviera, as it was informally called, was already developing, as many resorts, palaces and dachas began appearing along the south coast, particularly around Yalta and Alupka. Then, the sanatoria, or therapeutic resorts, appeared in many localities along the coast. By the turn of the century, the rather cold Russian Empire had already developed a warm "Mediterranean" destination of its own, albeit an exclusive one for nobility, the upper classes and the intelligentsia. During the Soviet period southern Crimea became a popular destination, not only for Communist Party officials, but also for workers. Even so, the idea of an exotic holiday destination within the Crimean peninsula had entered the imagination of the average Soviet citizen. It survives to this day, although now with a more commercial aspect.

Cultural Mediterraneanization in Crimea is also reflected in a number of other everyday aspects, beyond crops and tourism. Crimean cuisine, for example, has a strong Mediterranean element, mainly influenced by Turkish cuisine, as well as containing elements from Central Asian, Jewish, Russian and Ukrainian cuisines. Architecture is another aspect, apparent in many pre-Soviet buildings where Venetian, Moorish, Oriental and other styles are replicated. Dachas and many old houses in Yalta and Bakhshisaray have the type of curved tiles typical of the Mediterranean region.

Mediterranean-type land degradation?

As physical and cultural commonalities between southern Crimea and the Greater Mediterranean region become evident, one question emerges: Has southern Crimea been subject to the same type of long-term land degradation that the Mediterranean region underwent? In the search for an answer, it is important first to define what land degradation in the Mediterranean actually means.

Awareness of the degradation of Mediterranean lands and how to treat it begins with the major classical authors who wrote on the environment and agriculture, most notably Pliny the Elder, Collumella, Varro and Cato (Robinson 2006; Isager and Skydsgaard 1992; Hughes 1994). But the modern view of Mediterranean land degradation dates back to the landmark work of George Perkins Marsh, who described the destruction of forests and soil erosion so evident in the mid-1800s (Lowenthal 2000). Today, most texts dealing with Mediterranean environmental history include some of the ideas developed during the twentieth century regarding land degradation and conservation practices (e.g., Lowdermilk 1944; McNeill 1992; Hughes 2005; Vogiatzakis 2012). Popular and somewhat influential are those focusing on geomorphological and hydrological aspects (e.g., Vita Finzi 1969; Bottema, Entjes-Nieborg and van Zeist 1990; Lewin, Macklin, and Woodward 1995).

Most studies of environmental history offer definitions of what constitutes long-term and short-term Mediterranean land degradation. These definitions may be summarized thus:

> *Long-term Mediterranean land degradation* encompasses the millennial exposure to farming and pastoralism, which has degraded forests, creating secondary communities (i.e., maquis, garrigue and phrygana), and degraded soils exposing rock (mainly limestone and marls).The fluvial sediments from highly degraded environments then clog rivers and silt up estuaries. The process is aided by the pattern of a long dry season followed by a season of intense rain in a particularly mountainous terrain, which, in the context of vegetation removal, leads to a reduction in infiltration and an increase in runoff. Infiltration reduces the amount available to recharge aquifers while high runoff removes soil cover and surface sediments. Conservation measures vary, but the most common is agricultural terracing, a common feature of the cultural Mediterranean landscape.
>
> *Short-term Mediterranean land degradation* refers to the development of land transformation in the modern era (from the time of George Perkins Marsh in the early to mid-1800s), when peasant communities grew in numbers, leading to the widespread removal of trees and other vegetation, intensifying the problems of soil erosion and groundwater decline, and prompting some governments in the twentieth century to implement reforestation, which in many cases was successful. But as the century progressed, industrialization and rural emigration led to the abandonment of rural areas, stimulating growth of vegetation on previously terraced or badly damaged lands.

Does Crimea fit into the scheme described above? In terms of long-term degradation traits, southern Crimea seems to fit the general criteria of being a mountainous area with a millennial history of farming and pastoralism. The only difference with the Mediterranean region is that, in Crimea, Neolithization and overseas trade (e.g., Phoenician or Greek) came late in time, thus reducing the time the region was exposed to strong agricultural pressures. However, being part of large Mediterranean-based empires (e.g., Roman, Byzantine and Ottoman), and having at times large numbers of people practising viticulture, horticulture and pastoralism, does match the way things developed in the Mediterranean.

In terms of short-term land degradation, Crimea does not meet the description above, particularly so because during the past two decades it has been part of a different sphere of socio-economic influence, namely Russian and Soviet. This means that the developments described by George Perkins Marsh for the Mediterranean region do not tally with the situation in Crimea, where, at the time, different developments were occurring. In the 1800s, the native Turkic population of Crimea was gradually replaced by a number of groups, including Slavic and Germanic, many of whom had no tradition of developing mountain and Mediterranean type agriculture. Even more different has been the fact that the collectivization process dominated for most of the twentieth century—a development not seen in the Mediterranean areas, with the exception of the former Yugoslavia and Albania. The depopulation and decline of the rural Mediterranean did not occur in Crimea in the twentieth century. It seems, however, that in the twenty-first century, Crimea may follow the same path to Mediterranenization as several areas of the Mediterranean, particularly with regard to tourism, mimicking aspects of tourism in Turkey, Cyprus and other parts of the eastern Mediterranean.

Mediterraneanization, the economy and the environment

The scientific concept of Mediterraneanization began to develop at the beginning of the twentieth century and became well established during the Soviet period. But Mediterraneanized Crimea is more than just a concept. Mediterraneanization in Crimea has two sides: (1) a scientific side, driven mainly by studies of landscape evolution and conservation, and (2) a practical side, defined mainly in terms of being an economic asset that actually sells via certain products, particularly wine and tourism. The first began to develop in the early nineteenth century with the creation of the Nikitsky Botanical Gardens and the research that this institution catalyzed. The second emerged during the last two decades of the nineteenth century, particularly with the creation of the idea of a Russian Riviera. Eventually the two approaches began to merge, with the cultural and more tourism-oriented being the most important.

After the collapse of the Soviet Union, when state-sponsored vacations in Crimea disappeared, an industry began to develop, first for the rich (called the New Russians in the 1990s) and then for the middle class. However, unlike the rest of the Mediterranean, post-Soviet tourism in Crimea catered to visitors from Russia, Ukraine and other former Soviet republics. It was not until recent years that Crimea began to advertise (modestly) to a broader European public. One thing that did help was the removal of visa restriction to short-term visitors from Western countries, including the United States, during the presidency of Viktor Yushchenko. But the annexation of Crimea in March 2014 represented a setback to foreign tourism, including those from Ukraine. Economic sanctions, some of which target Crimea itself undermined tourism dramatically in the summer 2014.

Advertising a tourism destination as a Mediterranean experience but at a lower cost is understandable in our global era—it is an approach taken by many resorts, such as Ibiza, Crete, Korfu, Cyprus and the Aegean (Morris 2006). Yet although Crimea has benefited from its Mediterraneanization potential, this resource has not been fully exploited. If it is, it well may have a negative impact on the physical side of Mediterraneanization—i.e., the south coast's flora, fauna and their ecosystems, as areas for cultivation of commercial crops (e.g., grapes and essential oils) and tourism infrastructure are expanded. These prospective developments represent a dilemma and a challenge for the rational use of resources, an aspect that is now being brought to discussion in local scientific circles.

As for climate change, it is hard to tell what the landscape will be in the context of increasing global temperatures, since no study has specifically targeted vegetation and soils in southern Crimea. Evidence of local increases in temperatures are clear in the records of the twentieth century (Vyed' 2000; Yergina 2009), but no major changes in the ecosystems attributed to climate, as opposed to human development, have paralleled the temperature increases. Prospects for the wine industry show no clear development, because while warmth may improve certain varieties, other adverse climatic conditions that come with global warming may play a negative role (Korsakova 2011). Thus the cultural and physical resources sustained by Mediterraneanization deserve much more attention from the scientific community.

BIBLIOGRAPHY

Aksu, A.E., Mudie, P.J., Rochon, A., Kaminski, M.A., Abrajano, T. and Yaçar, D. 2002. "Persistent Holocene outflow from the Black Sea to the eastern Mediterranean contradicts Noah's Flood hypothesis". *GSA Today* 12: 4–10.

Aksyonov, V. 1983. *The Island of Crimea*, translated by M.H. Heim. New York: Random House.

Andrusov, N.I. 1912. *Terrasy Sudaka*. Kiev: Kiyevskoye Obshchestvo Yestyestvoyspitatelyey: Tipografija Imperatorskaya Universityeta.

Andryeyev, A.R. 1997. *Istoriya Kryma* [*History of Crimea*]. Moscow: Izdatyel'stvo Meyzhryegionalnyy Tsyentr Otraslevoy Informatiki Gosatomnadzora Rossii.

Apostolov, L.G., Bokov, V.A. and Dulitskiy, A.I. (eds). 1998. *Biologichyeskoye i Landshaftnoye raznoobraziye Kryma: Problyemy i pyerspyektivy* [*Biological and Landscape Diversity of Crimea: Problems and Perspectives*]. Simferopol: Tavriya.

Artyemyenko, I.I. 1985. *Arkhyeologiya Ukrainskoy SSR* [*Archaeology of the Ukrainian SSR*]. 3 vols. Kiev: Naukova Dumka.

Artyushyenko, A.T. and Mishnyov, V.G. 1978. *History of the vegetation on the yailas and surrounding slopes during the Holocene*. Kiev: Naukova Dumka.

Bagrova, L.A., Bokov, V.A. and Bagrov, N.V. 2001. *Geografiya Kryma* [*The Geography of Crimea*]. Kiev: Lybid'.

Balabanov, I.P. 2007. "Holocene sea-level changes of the Black Sea". In *The Black Sea Flood Question: Changes in Coastline, Climate and Human Settlement*, ed. V. Yanko-Hombach, A.S. Gilbert, N. Panin and P.M. Dolukhanov, 711–30. Berlin and New York: Springer.

Balandin, Yu. G. 1982. "Kolyebaniya urovniya moray v golotsyenye i ikh otrozhenii v stroyeniye litifitsirovannikh rakushechnykh otlozheniy Arabatskoy kosy Azovskoy Moriya" ["Holocene sea-level fluctuations as reflected in the lithified shell beaches of the Arabatskaya sand bar of the Azov Sea"]. In *Izmeniyeniya Urovniya Morya*, ed. Yu.G. Balandin and N.N. Trashchuk, 227–37. Moscow: Moskovskiy Gosudarstvyeniy Univyersityet.

Barg, I.M. and Ivanova, T.A. 2000. "Stratigrafiya i geologicheskoye razvitiye ravninogo Krymav Miotsyenye" ["Stratigraphy and geological development in the Crimean plains during the Miocene"]. *Stratigrafiya i Geologicheskaya Korryelyatsiya* 8: 83–93.

Baryshnikov, G. 2003. "*Mammuthus primigenius* from the Crimea and the Caucasus". In *Advances in Mammoth Research, Proceedings of the Second International Mammoth Conference, May 16–20, 1999*, DEINSEA 9, ed. J.W.F. De Vos and D. Mold, 41–56. Rotterdam: Natural History Museum.

——— 2006. "Late Pleistocene arctic fox (*Alopex lagopus*) from Crimea, Ukraine". *Quaternary International* 142–3: 208–17.

Bater, J.H. and French, R.A.R. (eds). 1983. *Studies in Russian Historical Geography*. 2 vols. London and Toronto: Academic Press.

Baykov, A.M. 2009. "Antropogyennoye zagryaznyeniye". In *Sovremenyie landshaftty Kryma i sopredel'nykh akvatoriy*, ed. Ye.A. Pozachyenyuk, 247–9; 319–21; 357–62; 390–1. Simferopol: Biznyes-Inform.

Bazhyenova, L.D. 1980. "Novyye materialy k izucheniyu vyurmskoy flory Gornogo Kryma" ["New materials for the study of the flora of the Crimean Mountains during the Würm glacial"]. *Tektonika i Stratigrafiya* (Kiev) 19: 101–3.

Behringer, W. 2010. *A Cultural History of Climate*. Cambridge, UK: Polity.

Berg, L.S. 1950. *Natural Regions of the USSR* (translated from the Russian by O.A. Titelbaum). New York: Macmillan.

Bibikov, S.N., Stanko, V.N. and Cohen, V.Yu. 1994. *Final'nyy Palolit i Mesolit gornogo Kryma*. [*The Final Paleolithic and Mesolithic of the Crimean Mountains*]. Odessa: Vyest'.

Bintliff, J.L. 2006. "Issues in the economic and ecological understanding of the chora of the classical polis in its social context: a view from the intensive survey tradition of the Greek homeland". In *Surveying the Greek Chora. The Black Sea Region in a Comparative Perspective*, ed. P.G. Bilde and V.F. Stolba, 13–26. Aarhus, Denmark: Aarhus University Press.

Biodiversity Support Program. 1999. "Priority-Setting in Conservation: A New Approach for Crimea". Washington DC: Biodiversity Support Program and World Wildlife Fund.

Blagovolin, N.S. 1975. "Soil relict and late Holocene tectonic movement in the North Crimean Lowland". In *Genesis and Geography of Soils*, ed. S.V. Zonn and V.M. Fridland, 232–44. New Dehli: Indian National Science Docent Centre.

Bonch-Osmolovski, G.A. 1925. "Early man in Crimea". *American Journal of Physical Anthropology* 8: 114.

Bondaryeva, L.V. 2008. "Flora i rastityel'nost' Geraklyeyskogo Poluostrova" ["Flora and vegetation of the Heraklean Peninsula"]. Dissertation for doctor of sciences. Yalta: Nikitsky Botanical Gardens – Scientific Research Center.

Borisyak, A.A. 1914. *Sevastopol'skaya fauna mlekopitayushikh* [*Mammal fauna of Sevastopol*]. Trudy Geologicheskogo Komiteta, New Series 87. St Petersburg: Geological Committee.

Bottema, S., Entjes-Nieborg, G. and van Zeist, W. (eds). 1990. *Man's Role in the Shaping of the Eastern Mediterranean Landscape*. Rotterdam: Balkema.

Bradley, R. 1725. *A Survey of Ancient Gardening and Husbandry*. London: B. Motte.

Braudel, F. 1971. *The Mediterranean and the Mediterranean World in the Age of Philip II*. Translated from the French by Siân Reynolds. New York: Harper and Row.

Browicz, K. and Zieliński, J. 1982–94. *Chorology of trees and shrubs in South-West Asia and adjacent regions*. 10 vols. Warsaw: Polish Academy of Sciences: Institute of Dendrology.

Bumbyer, K. Vagin, Yu, L.S., Klyepyenin, N.N. and Sokolov, V.V. (eds). 1913. *Krym: Putyevodityel'* [*Crimea: A Travel Guide*]. Simferopol: Tipografiya Tavrichyeskoy Gubyernii Zyemstva.

Burke, A. 2006. "Neanderthal settlement patterns in Crimea: A landscape approach". *Journal of Anthropological Archaeology* 25: 510–23.

Burroughs, W. 2005. *Climate Change in Prehistory*. Cambridge: Cambridge University Press.

Buynevich, I.V., Yanko-Hombach, A., Gilbert, A. and Martin, R.E. (eds). 2011. *Geology and Geoarchaeology of the Black Sea Region: Beyond the Flood Hypothesis*. Geological Society of America, Special Paper 473. Boulder, CO: Geological Society of America.

Byelyanskiy, I.L., Lyezina, I.N. and Supyeranskaya, A.V. 1998. *Krym: Geograficheskiye nazvaniya* [*Crimea: Geographic names*]. Simferopol: Bibliotyeka Krymoveda, Tavriya-Plyus.

Carter, J.C., Crawford, M., Lehman, M.P., Nikolaenko, G. and Trelogan, J. 2000. "The chora of Chersonesos in Crimea, Ukraine". *American Journal of Archaeology* 104: 707–41.

Carter, J.C. (ed.). 2003. *Chersonesos: City, Chora, Museum and Environs*. Austin: Institute of Classical Archaeology, University of Texas at Austin.

Chabai, V.P. 2005. "Kabazi II: Stratigraphy and archaeological sequence". In *Kabazi II: Last Interglacial. Occupation, Environment and Subsistence*, ed. V.P. Chabai, Richter, J. and T. Uthmeier, 1–24. Simferopol-Cologne: National Academy of Sciences of Ukraine, Institute of Archaeology, Crimean Branch, and University of Cologne, Institute of Prehistoric and Protohistoric Archaeology.

——— 2007. "The Middle Paleolithic and Early Upper Paleolithic in the northern Black Sea Region". In *The Black Sea Flood Question: Changes in Coastline, Climate and Human Settlement*, ed. V. Yanko-Hombach, A.S. Gilbert, N. Panin and P.M. Dolukhanov, 279–97. Berlin and New York: Springer.

Chabai, V.P., Marks, A.E. and Monigal, K. 1999. "Western Crimea Middle Paleolithic paleoenvironment and paleoeconomy". In *The Middle Paleolithic of Western Crimea, Vol. 2*, Etudes et Recherches Archéologiques de l'Université de Liège 84, ed. V.P. Chabai and K. Monigal, 211–33. Liège: University of Liège.

Chabai, V.P., Monigal, K. and Marks, A.E. 2004. "Crimea in the context of the eastern European Middle Paleolithic and Early Upper Paleolithic". In *The Middle Paleolithic and Early Upper Paleolithic of Eastern Crimea*, Etudes et Recherches Archéologiques de l'Université de Liège 104, ed. V.P. Chabai, K. Monigal and A.E. Marks, 419–60. Liège: University of Liège.

Chashchin, A.K. 1996. "The populations Black Sea populations of anchovy". *Scientia Marina* 60: 219–25.

Chepalyga, A.L. 2007. "The Late Glacial Great Flood in the Ponto-Caspian Basin". In *The Black Sea Flood Question: Changes in Coastline, Climate and Human Settlement*, ed. V. Yanko-Hombach, A.S. Gilbert, N. Panin and P.M. Dolukhanov, 119–48. Berlin and New York: Springer.

Çelebi [=Chelebi], E. 1996. *Book of Travels: March with the Tatars and Travels in Crimea* (1641–7). Simferopol: Tavriya.

Çelebi, E. and von Hammer-Purgstall, J. 1834. *Narrative of Travels in Europe, Asia and Africa in the Seventeenth Century,* by Evliya Efendi, translated from the Turkish by the Ritter Joseph von Hammer. London: Parbury Allen and Co.

Cohen, V. 1996. "Neolithization of the Crimean Mountains: current stage of investigations". *Préhistoire Européenne* 9: 417–33.

——— 2000–1. "Landscape, economy and complexity in light of the Crimean Final Paleolithic and Mesolithic data: preliminary analysis". *Préhistoire Européenne* 16–17: 231–54.

Coleman, D.F. and Ballard, R.D. 2007. "Submerged paleoshorelines in the southern and western Black Sea – implications for inundated prehistoric archaeological sites". In *The Black Sea Flood Question: Changes in Coastline, Climate and Human Settlement*, ed. V. Yanko-Hombach, A.S. Gilbert, N. Panin and P.M. Dolukhanov, 671–96. Berlin and New York: Springer.

Comes, H.P. and Kadereit, J.W. 1998. "The effect of Quaternary climatic changes on plant distribution". *Trends in Plant Sciences* 3: 432–8.

Conkle, M.T., Schiller, G. and Grünwald, C. 1988. "Electrophoretic analysis of diversity and phylogeny of Pinus brutia and closely related taxa". *Systematic Botany* 13: 411–24.

Cordova, C.E. 2007. "Holocene Mediterranization of the Southern Crimean vegetation: paleoecological records, regional climate change, and possible non-climatic influences". In *The Black Sea Flood Question: Changes in Coastline, Climate and Human Settlement*, ed. V. Yanko-Hombach, A.S. Gilbert, N. Panin, and P.M. Dolukhanov, 319–44. Berlin and New York: Springer.

Cordova, C.E. and Lehman, P.H. 2003. "Achaeopalynology of synanthropic vegetation in the Chora of Chersonesos, Ukraine". *Journal of Archaeological Science* 30: 1483–501.

——— 2005. "Holocene environmental change in Southwestern Crimea (Ukraine) in pollen and soil records". *The Holocene* 15: 263–77.

——— 2006. "Mediterranean agriculture in southwestern Crimea: paleoenvironments and early adaptations". In *Beyond the Steppe and the Sown*, Colloquia Pontica 13, ed. D.L. Peterson, L.M. Popova and A.T. Smith, 425–47. Leiden: Brill

Cordova, C.E., Gerasimenko, N., Lehman, P. and Kliukin [=Klyukin], A.A. 2011. "Late Pleistocene and Holocene paleoenvironments of Crimea: pollen, soils, geomorphology and geoarchaeology". In *Geology and Geoarchaeology of the Black Sea Region: Beyond the Flood Hypothesis*, Geological Society of America, Special Paper 473, ed. I. Buynevich, V. Yanko-Hombach, A. Gilbert and R.E. Martin, 133–64. Boulder, CO: Geological Society of America.

Cordova, C.E., Rybak, A. and Lehman, P.H. 2001. "Vegetation Patterns and Conservation Issues in Southern Crimea". *Post-Soviet Geography and Economics* 42: 49–72.

Daget, P. and David, P. 1982. "Essai de comparaison des diverses approches climatiques de la méditerranéité". *Ecologia Mediterranea* 8: 33–48.

Davis, P.H. 1982. *Flora of Turkey and the East Aegean Islands*, vol. 7. Edinburgh: Edinburgh University Press.

Demidenko, Yu. E. 2008. "The Early and Mid-Upper Palaeolithic of the North Black Sea Region: An overview". *Quartär* 55: 99–114.

Didukh, Ya.P. 1985. "Sistematika i istoriya razvitiya buka i bukovykh lyesov gornogo Kryma". ["Systematics and history of beech and beech forests in the Crimean Mountains"]. *Botanichyeskiy Zhurnal* 8: 1040–50.

——— 1992. *Rastityel'ny pokrov gornogo Kryma*. [*The vegetation cover of the Crimean Mountains*]. Kiev: Naukova Dumka.

——— 2003. "Rastitel'nost'" ["Vegetation"]. In *Atlas: Avtonomnaya Respubklika Krym* [*Atlas: the Autonomous Republic of Crimea*], ed. M.B. Bagrov and L.G. Rudenko, 32. Kiev and Simferopol: Tavrida National University, Crimean Center of the National Ukrainian Academy of Sciences, Institute of Geography of the NAN, and Institute for Transfer Technology.

Didukh, Ya.P., Litvinskaya, S.A. and Novosad, V.V. 1990. "Botanico-geographical regionalization of the Crimea-Novorossiysk Province". *Botanicheskiy Zhurnal* 75: 494–507.

Dogushev, I.A. 1979. "North Crimean Canal". In *Great Soviet Encyclopaedia*, vol. 23, translation of the 3rd edn, ed. A.M. Prokorov, 168–9. New York: Macmillan.

Douguedroit, A. and Zimina, R.P. 1987. "Le climat méditerranéen en France et en U.R.S.S." *Méditerranée* 61: 75–84.

Dragan N.A. 2003. "Pochvy" ["Soils"]. In *Atlas: Avtonomnaya Respubklika Krym* [*Atlas: the Autonomous Republic of Crimea*], ed. M.B. Bagrov and L.G. Rudenko, 30. Kiev and Simferopol: Tavrida National University, Crimean Center of the National Ukrainian Academy of Sciences, Institute of Geography of the NAN, and Institute for Transfer Technology.

——— 2004. *Pochvenniye resursy Kryma*. Simferopol: Doliya.

——— 2009. "Pochvy" ["Soils"]. In *Sovremenyie landshaftty Kryma i sopredel'nykh akvatoriy*, ed. Ye.A. Pozachyenyuk, 88–128. Simferopol: Biznyes-Inform.

Druzhinina, Ye. I. 1959. *Syevyernoye Prichyernomor'ye v 1775–1800 gg* [*The northern Black Sea Region in the years 1775–1800*]. Moscow: Akadyemiya Nauk SSSR, Institut Istorii.

Dubois de Montpéreux, F. 1843. *Voyage autour du Caucase, chez les tcherkesses et les abkhases, en Colchide, en Géorgie, en Arménie et en Crimée* (5 vols). Paris: Librairie de Gide.

Dulitskiy, A.I. 2001. *Mlekopitajushchiye: Istoriya, sostayaniye, okhrana, perspektivy* [*Mammals: history, status, conservation, perspectives*]. Simferopol: SONAT.

Dyulichyev, V.P. 1997. *Rasskazy po istorii Kryma* [*Accounts of Crimean History*]. Simferopol: Biznyes-Inform.

Dzyens-Litovskaya, N.N. 1970. *Pochvy i rastitel'nost' stepnogo Kryma* [*Soils and vegetation of the Crimean steppe*]. Leningrad (St Petersburg): Nauka.

Eckenwalder, J.E. 2009. *Conifers of the World*. Portland, OR: Timber Press.

Emberger, L. 1955. "Une classification biogéographique des climats". *Naturalia Monspeliaca, Série Botanique* 7: 3–43.

Faber, P.L. 2000. *Finding Order in Nature: The Naturalist Tradition from Linnaeus to E.O. Wilson*. Baltimore and London: Johns Hopkins.

Fagan, B.M. 2010. *Cro-Magnon: How the Ice Age Gave Birth to the First Modern Humans*. New York, NY: Bloomsbury.

Farjon, A. 2001. *World Checklist and Bibliography of Conifers*. London: Kew Royal Botanical Gardens.

Finlayson, C. 2004. *Neanderthals and Modern Humans: An Ecological and Evolutionary Perspective*. Cambridge: Cambridge University Press.

Fisher, A. 1978. *The Crimean Tatars*. Stanford, CA: Hoover Institution Press.

Frolov, V.T. 1998. "Mezozoyskiye i kaynozoyskiye formatsii Kryma (geneticheskiy analiz)" ["Mesozoic and Cenozoic formations of Crimea (analysis of their origin)"]. Byuletyen' *Byullyetyen' Moskovskogo Obshchestva Ispytatyeley Prirody*, Otdyel Geologii 73: 39–48.

Fursa, D.I. 1977. *Pogoda, orosheniye i produktivnost' vinograda* [*Weather, irrigation and grape productivity*]. Leningrad (St Petersburg): Gidrometeoizdat.

Fyodorov, P.V. 1978. *Plyeystotsyen Ponto-Kaspiya*. [*The Pleistocene of the Black Sea and Caspian Seas*]. Moscow: Nauka.

Garkusha, L.Ya. 1998. "Raznoobraziye lyesov Gornogo Kryma" ["Forest diversity of the Crimean Mountains"]. In *Biologicheskoye i Landshaftnoye raznoobraziye Kryma: Problemy i perspektivy*, ed. L.G. Apostolov, V.A. Bokov and A.I. Dulitskiy, 84–8. Simferopol: Tavriya.

Gavrilov, A.V. 1997. "Environmental and climatic changes between 3 and 5 ka BP in Southeastern Ukraine". In *Third Millennium BC Climate Change and Old World Collapse*, ed. H.N. Dalfes, G. Kukla and H. Weiss, 371–99. Berlin Heidelberg: Springer Verlag.

——— 2006. "Theodosia and its chora in Antiquity". In *Surveying the Greek Chora. The Black Sea Region in a comparative perspective*, ed. P.G. Bilde and V.F. Stolba, 249–72. Aarhus, Denmark: Aarhus University Press.

Gerasimenko, N. 1997. "Environmental and climatic changes between 3 and 5 ka BP in Southeastern Ukraine". In *Third Millennium BC Climate Change and Old World Collapse*, ed. H.N. Dalfes, G. Kukla, and H. Weiss, 371–99. Berlin Heidelberg: Springer Verlag.

——— 2004. "Vegetational history of Buran Kaya III". In *The Middle Paleolithic and Early Upper Paleolithic of Eastern Crimea*, Etudes et Recherches Archéologiques de l'Université de Liège, vol. 104, ed. V. P. Chabai, K. Monigal, and A. E. Marks, 19-34. Liège: University of Liège.

——— 2006. "Upper Pleistocene loess-paleosol and vegetational successions in the Middle Dnieper Area, Ukraine". *Quaternary International* 149: 55–66.

——— 2007. "Environmental changes in the Crimean Mountains during the Last Interglacial-Middle Pleniglacial as recorded by pollen and lithopedology". *Quaternary International* 164–5: 207–20.

——— 2011. "Climatic and environmental oscillations in southwestern Ukraine from 30 to 10 ka, inferred from pollen and lithopedology". In *Geology and Geoarchaeology of the Black Sea Region: Beyond the Flood Hypothesis, Geological Society of America*, Special Paper 473, ed. I. Buynevich, V. Yanko-Hombach, A. Gilbert, and R.E. Martin, 117–32. Boulder, CO: Geological Society of America.

Gerasimenko, N., Subetto, D., Bakhmutov, V. and Dubis, L. 2011. "Late Middle Holocene climatic oscillations recorded in pollen from annual laminated deposits of Saki Lake (Ukraine)". In *Learning from Other Communities: Towards More Robust Varve Chronologies. Program with Abstracts*, 41–3. Corpus Christi, USA: PAGES Varves Working Group, INQUA Project Number 1102.

Gerasimov, I.P. 1954. *Korichnevye pochvy sredizemnomorskikh oblastyey* [*Cinnamonic soils of the Mediterranean Regions*]. Lecture in the 5th International Congress of Pedologists: Moscow, Izdatel'stvo Akademii Nauk SSSR.

Glebov, A.Yu. and Shel'ting, S.K. 2007. "Sea-level changes and coastline migrations in the Russian sector of the Black Sea: Applications to the Noah's flood hypothesis". In *The Black Sea Flood Question: Changes in Coastline, Climate and Human Settlement*, ed. V. Yanko-Hombach, A.S. Gilbert, N. Panin and P.M. Dolukhanov, 631–773. Berlin, New York: Springer.

Golubyev, V.N. 1996a. *Biologicheskaya Flora Kryma* [*Biological Flora of Crimea*]. 2nd edn. Yalta: National Nikitsky Botanical Gardens.

——— 1996b. "K istorii formirovaniya flory Kryma" ["Towards a history of the formation of Crimea's flora"]. *Byulletyen' Nikitskogo Botanichyeskogo Sada* 61: 5–9.

——— 1997. "O genezisye biomrfologichyeskikh tipov vegetatsii i perezimovki rastyeniy reliktovoy sredisyemnomorskoy rastityelnosti yuzhnogo berega Kryma" ["On the genesis of biomorphological types of vegetation and winter resistance of plants of the relict Mediterranean vegetation of the South Coast of Crimea"]. In *Struktura Flora i rastitel'nosti Kryma*, ed. V.V. Korzhyenyevskiy, 7–21. Yalta: National Ukrainian Academy of Sciences-State Nikitsky Botanical Gardens.

——— 1998. "Gyeografichyeskaya struktura flory Kryma" ["Geographical Structure of the Crimean Flora"]. In *Biologicheskoye i Landshaftnoye raznoobraziye Kryma: Problemy i perspektivy*, ed. L.G. Apostolov, V.A. Bokov and A.I. Dulitskiy, 66–7. Simferopol: Tavriya.

Gömöry, D., Paule, L. and Vyšný, J. 2007. "Patterns of allozyme variation in western Eurasian Fagus". *Botanical journal of the Linnean Society* 154: 165–74.

Goncharenko, G.G., Bolsun, S.I., Nyevo, E. and Zakhavi, A. 1998. "Genetiko-taksonomicheskiye vsaimootnoshyeniya u Sosny Pitsundskoy, Sosny stankevicha i Sosny Brutskoy" ["Genetic and taxonomical relations between *Pinus pityusa*, *Pinus stankewiczii* and *Pinus Brutia*"] *Doklady Rossiyskoy Akademii Nauk* 359: 565–8.

Gordina, A.D., Pavlova, E.V., Ovsyany, E.I., Wilson, J.G., Kemp, R.B., and Romanov, A.S. 2001. "Long-term changes in Sevastopol Bay (the Black Sea) with particular reference to the ichthyoplankton and zooplankton". *Estuarine, Coastal and Shelf Science* 52: 1–13.

Gorodyetsky, A.A. and Karpyenko, S.A. 2000. "Ostonvanyye podkhody ryealizatsiya proyekta national'nogo prirodnogo parka 'Sivashkiy'" ["The main approaches to the implementation of the project to establish the 'Sivashki' National Nature Park"]. In *Sovryemyennoye Sostoyaniye Sivasha: Sbornik Nauchnikh Statyey*, ed. G.A. Gorodiskaya, 93–104. Kiev: Wetlands International.

Gozhik, P., Matviishina, Zh., Gerasimenko, N., Rekovets, L. and Shelkoplyas, V. 2001. "Quaternary stratigraphy". In *Proceedings of the Conference of Subcommission on European Quaternary Stratigraphy. The Quaternary of Ukraine Explored. Excursion Guide of the SEQS 2001 Conference*, 8–11. Kiev. Subcommission on European Quaternary Stratigraphy.

Gozhik, P.F. and Novosyel'skiy, F.A. 1989. *Gyeologichyeskiye usloviya stroityel'stva Dneprovsko-Bugskogo gidrouzla* [*The geological conditions of the Dnieper-Bug waterworks*]. Kiev: Naukova Dumka.

Gratsianskiy, A.N. 1971. *Priroda Sryedizyemnomor'ya* [*The Nature of the Mediterranean*]. Moscow: Izdatel'stvo Mysl'.

Greipsson, S. 2011. *Restoration Ecology*. Sudbury, Massachussetts: Jones and Bartlett Publishers.

Grosset, G.E. 1979a. "O proiskhozhdyenii flora Kryma. Soobshchyeniye 2" ["On the origin of the flora of Crimea, Communication 2"]. *Byullyetyen' Moskovskogo Obshchestva Ispytatyeley Prirody, Otdyel Biologichyeskiy* 84 (1): 64–84.

——— 1979b. "O proiskhozhdyenii flora Kryma. Soobshchyeniye 2" ["On the origin of the flora of Crimea, Comunication 2"]. *Byullyetyen' Moskovskogo Obshchestva Ispytatyeley Prirody, Otdyel Biologichyeskiy* 84 (2): 35–55.

Grove, J.M. and Conterio, A. 1994. "Climate in the eastern and central Mediterranean, 1675 to 1715". In *Climatic Trends and Anomalies in Europe, 1675 to 1715*, ed. B. Frenzel, C. Pfister and B. Glaeser, 275–86. Stuttgart: Fischer.

Gyertsyen, A.G. 2003. "Sryednyevyekovoy Krym (V-XIII vv.)" ["Medieval Crimea (5th–13th centuries)"]. In *Atlas: Avtonomnaya Respubklika Krym* [*Atlas: The Autonomous Republic of Crimea*], ed. M.B. Bagrov and L.G. Rudenko, 12. Kiev and Simferopol: Tavrida National University, Crimean Center of the National Ukrainian Academy of Sciences, Institute of Geography of the NAN, and Institute for Transfer Technology.

——— 2010. "Khazary" ["The Khazars"]. In *Ot Kimyeriytsiev do Krymchakov*, ed. I.N. Khrapunov and A.G. Gyertsyen, 140–61. Simferopol: Dolya.

Habliz [Hablits], K.I. 1789. *Physicalische Beschreibung der Taurischen Statthalterschaft nach ihrer Lage und allen drei Naturreichen* (translated from the Russian by L. Guckenberger). Hannover and Osnabruck: Verlage der Schmidtschen Buchhandlung.

Halenko, O. 2004. "Wine production, marketing and consumption in the Ottoman Crimea, 1520–1542". *Journal of the Economic and Social History of the Orient* 47: 507–47.

Hannestad, L. 2007. "Timber as a trade resource of the Black Sea". In *The Black Sea in Antiquity: Regional and Interregional Economic Exchanges*, ed. V. Gabrielsen and J. Lund, 85–100. Aarhus, Denmark: Aarhus University Press.

Hansen, V.D. 1999. *The Other Greeks: The Family Farm and the Agrarian Roots of Western Civilization*. 2nd edn. Berkeley and Los Angeles: University of California Press.

Hazel, J. 2002. *Who's Who in the Greek World*. London: Routledge.

Hellegers, P.J. and Perry, C.J. 2006. "Can irrigation water use be guided by market forces? Theory and practice". *Water Resources Development* 22: 79–86.

Hempel, L. 1987. "The 'Mediterraneanization' of the climate in Mediterranean countries: a cause of the unstable ecobudget". *Geojournal* 14 (2): 163–73.

Hind, J. 1998. "Megarian Colonisation in the western Half of the Black Sea (Sister-and Daughter-Cities of Herakleia)". In *The Greek Colonisation of the Black Sea Area*, ed. G.R. Tsetskhladze, 131–52. Stuttgart: Franz Steiner Verlag.

Huggett, R.J. 2004. *Fundamentals of Biogeography*, 2nd edn. Routledge.

Hughes, J.D. 1994. *Pan's Travail: Environmental Problems of the Ancient Greeks and Romans*. Baltimore and London: The Johns Hopkins University Press.

——— 2005. *The Mediterranean: An Environmental History*. Santa Barbara, CA: ABC-CLIO.

——— 2006. *What is Environmental History?* Cambridge, UK: Polity Press.

Isager, S. and Skydsgaard, J.E. 1992. *Ancient Greek Agriculture: An Introduction*. London: Routledge.

Janushevich, Z. and Nikolaenko, G. 1979. "Fossil remains of cultivated plants in the ancient Tauric Chersonesos". In *Festschrift Maria Hopf*, Archaeo Physika 8, ed. U. Korber-Grohne, 115–34. Cologne: Rheinland Verlag.

Josephson, P.R. 2013. *An Environmental History of Russia*. Cambridge: Cambridge University Press.

Kalynychenko, O.V. 2003. "Rarytetni ridkolissya pivdeno-zakhidnoho uzberezhzhya Herakleys'koho Pivostrova" ["Rare species of the woodlands on the southwest coast of the Heraklean Peninsula"]. *Ukrayinsky Botanichny Zhurnal* 60: 652–8.

Karpyenko, S.A. and Lychak, A.I. 2000. "Perspyektivy rekreatsyionnogo razvitiya krymskogo Prisivashiya" ["Prospects for the development of recreational potential in areas of the Crimea adjacent to the Sivash Gulf"]. In *Sovryemyennoye Sostoyaniye Sivasha: Sbornik Nauchnikh Statyey*, ed. G.A. Gorodiskaya, 86–92. Kiev: Wetlands International.

Karpyenko, S.A., Yermakov, Yu. A., Glushchyenko, I.V. and Vatsyet, Ye.Ye. 2003. "Prirodno-zapovyednyye territorii" ["Protected natural territories"]. In *Atlas: Avtonomnaya Respubklika Krym* [*Atlas: The Autonomous Republic of Crimea*], ed. M.B. Bagrov and L.G. Rudenko, 75. Kiev and Simferopol: Tavrida National University, Crimean Center of the National Ukrainian Academy of Sciences, Institute of Geography of the NAN, and Institute for Transfer Technology.

Kelterbaum, D., Brückner, H., Dikarev, V., Gerhard, S., Pint, A. and Porotov, A. 2012. *Geoarchaeology* 27: 206–19.

Kholodov, V.N. 2002. "Mud volcanism: distribution, regularities and genesis. Communication 1: mud volcanic provinces and morphology of mud volcanoes". *Lithology and Mineral Resources* 37: 197–209.

King, C. 2004. *The Black Sea: A History*. New York: Oxford University Press.

King, R. 1997. "Introduction: An essay on Mediterraneanism". In *The Mediterranean Environment and Society*, ed. R. King, L. Proudfoot and B. Smith, 1–11. London: Arnold.

Klyukin, A.A. 2005. "Ekctryemal'nyye proyavlyeniya nyeblagopriyatnykh I opasnykh ekzogyennykh protsyessov v XX vyekye v Krymu" ["Extreme manifestations of adverse and dangerous exogenic processes during the 20th century in Crimea"]. *Gyeopolitika i ekogyeodynamika ryegionov* 1: 27–38.

Klyukin, A.A. and Korzhyenyevskiy, V.V. 2008. *Krymskoye Priazov'ye* [*The Azov region of Crimea*]. Simferopol: Biznyes-Inform.

Klyukin, A.A. and Shchyepinskiy, A.A. 1983. "Novyye palyeolitichyeskiye nakhodki v otlozhyeniyakh terras r. Al'ma (Krym)" ["New Paleolithic findings in the terrace deposits of the Al'ma River (Crimea)"]. *Biullyetyen' Komissii po Izucheniiu Chyetvyortichnogo Perioda* 52: 156–64.

Kobyechinskaya, V.G., Svolynskiy, A.D., Svolynksiyj, M.D. and Kapitonov, V.V. 2010. "Vedushchiye antropogennye factory narushayushchie stabil'nost' ekosityem Yaltinskogo gorno-lyesnogo prirodnogo zapovyednika" ["Leading anthropogenic factors in the disruption of ecosystem stability in the Yalta Forest Mountain Reserve"]. *Ekosistemy, ikh Optimizatsiya i Okhrana* 2: 58–74.

Koch, C.W. 1855. *The Crimea: With a Visit to Odessa*. London: George Routledge & Co.

Kochkin, M.A. 1967. *Pochvy, lyesa i klimat Gornogo Kryma i puti ikh ratsional'nogo ispolzovaniya* [*Soils, forests and climate of the Crimean Mountains and ways to their rational use*]. Moscow: Kolos.

Kolosov, Yu. G., Kharitonov, V.M. and Yakimov, V.P. 1975. "Paleoanthropic specimens from the site Zaskalnaya VI in Crimea". In *Paleoanthropology: Morphology and Paleoecology*, ed. R. Tuttle, 419–28. The Hague: Mouton.

Kolosov, Yu. G.V., Stepanchuk, N. and Chabai, V.P. 1993. *Ranniy Palyeolit Kryma [The Early Paleolithic of Crimea]*. Kiev: Naukova Dumka.

Korsakova, S. 2011. "Impact of climate change on the grape productivity in the southern coast of the Crimea". In *Challenges and Opportunities in Agrometeorology*, ed. S.D. Attri, L.S. Rathore, M.V.K. Sivakumar and S.K. Dash, 385–96. Berlin Heidelberg: Springer.

Kosarev, A.A. and Kostianoy, A.G. 2008. "Introduction". In *The Black Sea Environment. Handbook of Environmental Chemistry, Part 5*, ed. A.N. Kosarev, 1–10. Berlin: Springer Verlag.

Kosarev, A.N., Arkhipkin, V.S. and Surkova, G.V. 2008. "Hydrometeorological conditions". In *The Black Sea Environment. Handbook of Environmental Chemistry, Part 5*, ed. A.N. Kosarev, 135–58. Berlin: Springer Verlag.

Kostin, S.Yu. 1983. *Ptitsy Kryma [The birds of Crimea]*. Moscow: Nauka.

——— 1998. "Krasnaya Kniga flory i fauny Kryma" ["The Red Book of Crimean flora and fauna"]. In *Biologichyeskoye i landshaftnoye raznoobraziye Kryma: problemy i perspektivy*, ed. L.G. Apostolov, V.A. Bokov and A.I. Dulitskiy, 140. Simferopol: Tavriya.

Kostin, S.Yu. and Dulitskiy, A.I. 1998. "Sovryemyeniye sostayaniye genofonda fauny Kryma" ["Present status of faunal genetic pool in Crimea"]. In *Biologicheskoye i landshaftnoye raznoobraziye Kryma: problemy i perspektivy*, ed. L.G. Apostolov, V.A. Bokov and A.I. Dulitskiy, 143–5. Simferopol: Tavriya.

Kostopolou, E. and Jones, P.D. 2007. "Comprehensive analysis of the climate variability in the eastern Mediterranean. Part II: relations between atmospheric circulation patterns and surface climatic events". *International Journal of Climatology* 27: 1351–71.

Kravchenko, Ye. A. 2004. "Materialy doby pizn'oyi bronzy z poselennya Uch-Bash" ["Late Bronze materials from the Uch-Bash settlement"]. *Arkheologiya* 4: 52–66 (in Ukrainian).

Kremenetski, C.V. 1995. "Holocene vegetation and climate history of southwestern Ukraine", *Review of Palaeobotany and Palynology* 85: 289–301.

Krupskiy, N.K. and Polupan, N.I. 1979. *Atlas pochv Ukrainskoy SSR [Atlas of soils of the Ukrainian SSR]*. Kiev: Urozhay.

Kryukova, I.V., Zheribor, V.I., Kostin, YuV. and Kostin S.Yu. 1998. *Ryedkiye rastyeniya i zhivotnye Kryma [Rare plants and animals of Crimea]*. Simferopol: Tavriya.

Kudelsky, A.V. 2011. "Global geoenvironmental problems: Black Sea Basin". *Water Resources* 38: 849–58.

Kuprin, P.N. and Sorokin, V.M. 2007. "On the post-glacial changes in the level of the Black Sea". In *The Black Sea Flood Question: Changes in Coastline, Climate and Human Settlement*, ed. V. Yanko-Hombach, A.S. Gilbert, N. Panin and P.M. Dolukhanov, 205–20. Berlin and New York: Springer.

Kutajsov, V.V. 2006. "The chora of Kerkinitis". In *Surveying the Greek Chora. The Black Sea Region in a Comparative Perspective*, ed. P.G. Bilde and V.F. Stolba, 141–50. Aarhus, Denmark: Aarhus University Press.

Lazukov, G.I., Gvozdover, M.D., Roginskiy, YaYa, Uryson, M.I., Kharitonov, M.K. and Yakimov, V.P. 1981. *Priroda I dryevniy chyelovyek: Osnovnnyye etapy razvitiya prirody, palyeoliticheskogo chyelovyeka I yego kul'tury ha territorii SSSR v Plyeystotsyene. [Nature and prehistoric people: Basic phases of development of prehistoric peoples and their culture in the territory of the USSR during the Pleistocene]* Moscow: Mysl.

Le Houérou, H.N. 2005. *The Isoclimatic Mediterranean Biomes: Bioclimatology, Diversity and Phytogeography*. Montpellier, France: Henry Noel Le Houérou (2 vols).

Leroy, S.A.G. and Arpe, K. 2007. "Glacial refugia for summer-green trees in Europe and south-west Asia as proposed by ECHAM3 time-slice atmospheric model simulations". *Journal of Biogeography* 34: 2, 115–28.

Lewin, J., Macklin, M.G. and Woodward, J.C. (eds). 1995. *Mediterranean Quaternary River Environments*. Rotterdam, The Netherlands: A.A. Balkema.

Lisetskii, F.N. and Ergina [=Yergina], E.I. 2010. "Soil development in the Crimean Peninsula during the Holocene". *Eurasian Soil Science* 43: 601–3.

Lisetskii, F.N., Stolba, V.F., Ergina [=Yergina], E.I., Rodionova, M.E. and Terekhin, E.A. 2012. "Post-agrogenic evolution of soils in ancient Greek land use areas in the Herakleian Peninsula, southwestern Crimea". *The Holocene* 23: 504–14.

Lowdermilk, W.C. 1944. *Palestine: Land of Promise*. New York: Harper and Brothers Publishers.

Lowenthal, D. 2000. *George Perkins Marsh: Prophet of Conservation.* Seattle: University of Washington Press.

Lyebyedinskiy, V.I. 1982. *S gyeologichyeskim molotkom po-Krymu [With a geological hammer in Crimea]*. Moscow: Nyedra.

Lyevkovskaya, G.M. 1970. "Rekonstruksiya palyeogeograficheskikh usloviy gorodishcha Chayka po dannym sporovo-pyl'tsyevogo analiza analiza" ["Reconstruction of paleo-gegographic conditions of the Chayka site using pollen-spore analysis"]. *Kratkiye Soobshchenyeniya Instituta Arkhyeologii AN SSSR (KSIA)* 124: 102–8.

MacDonald, G., 2003. *Biogeography: Space, Time and Life.* New York: Wiley and Sons.

Malyeyev, V.P. 1946. "Osnovnyye etapy razvitya rastityel'nosti Sredizyemnomor'ya v chyetvyortichny period v sviazi s istoryey rastitel'nosti yuga SSSR". *Trudy Instituta Gyeografii* 37: 321.

——— 1948a. "Osnovnyye etapy razvitya rastityel'nosti Sredizyemnomor'ya I gornykh oblastyey yuga SSSR (Kavkaz I Kryma) v chyetvyortichny period". *Trudy Gosudarstvyennogo Nikitskogo botanichyeskogo sada imyeny V.M. Molotova* 25: 3–28.

——— 1948b. "Rastityel'nost' yuzhnogo Kryma". *Trudy Gosudarstvyennogo Nikitskogo botanichyeskogo sada imyeny V.M. Molotova* 25: 29–48.

Markova, A.K. 2000. "The Mikulino (= Eemian) mammal faunas of the Russian Plain and Crimea". *Geologie en Mijnbow/Netherlands Journal of Geosciences* 79: 293–301.

——— 2005. "Small mammals from the Paleolithic site of Kabazi II, western Crimea". In *Kabazi II: Last Interglacial Occupation, Environment and Subsistence. The Palaeolithic Sites of Crimea. Volume 1*, ed. V. Chabai, J. Richter and T. Uthmeier, 51–65. Simferopol-Cologne: National Academy of Sciences of Ukraine, Crimean Filial and Institute of Prehistory and Protohistory, University of Cologne.

Markova, A.K. and Puzachenko, A. 2006. "Vertebrate records: Late Pleistocene of Northern Asia". In *Encyclopedia of the Quaternary Period*, vol. 4, ed. E. Scott, 3158–75. Amsterdam: Elsevier Science.

Markova, A.K., Simakova, A.N. and Puzachenko, A.Yu. 2009. "Ecosystems of Eastern Europe at the time of maximum cooling of the Valdai glaciation (24–18 kyr BP) inferred from data on plant communities and animal assemblages". *Quaternary International* 201: 53–9.

Martin, R.E. and Yanko-Hombach, V. 2011. "Rapid Holocene sea-level and climate change in the Black Sea: An evaluation of the Balabanov sea-level curve". In *Geology and Geoarchaeology of the Black Sea Region: Beyond the Flood Hypothesis*, Geological Society of America, Special Paper 473, ed. I. Buynevich, V. Yanko-Hombach, A. Gilbert and R.E. Martin, 51–8. Boulder, CO: Geological Society of America.

Martin, T.E. 1996. *Ancient Greece: From Prehistoric to Hellenistic Times.* New Haven: Yale University Press.

Maslyennikov, A.A. 1997. "Isslyedovaniye syel'skoy tyerritorii yevropyeyskogo Bospora: itogi i pyeprektivy k kontsu vyeka" ["Research on the rural territories of the European Bosphorus: results and perspective for the end of the century"]. *Rossiyskaya Arkhyeologiya* 3: 60–70.

Matoshko, A., Gozhik, P. and Semenenko, V. 2009. "Late Cenozoic fluvial development within the Sea of Azov and Black Sea coastal plains". *Global and Planetary Change* 68: 270–87.

McCormick, M., Büntgen, U., Cane, M.A., Cook, E.R., Harper, K., Huybers, P., Litt, T., Manning, S.W., Mayewski, P.A., More, A.F.M., Nicolussi, K. and Tegel, W. 2012. "Climate change during and after the Roman Empire: Reconstructing the past from scientific and historical evidence". *Journal of Interdisciplinary History* 43: 169–220.

McNeill, J.R. 1992. *The Mountains of the Mediterranean World: An Environmental History.* New York: Cambridge University Press.

Meijers, M.J.M., Vrouwe, B., van Hinsbergen, D.J.J., Kuiper, K.F., Wijbrans, J., Davies, G.R., Stephenson, R.A., Kaymakci, N., Matenco, L. and Saintot, A. 2010. "Jurassic arc

volcanism in Crimea (Ukraine): implications for the paleo-subduction zone configuration of the Black Sea Region". *Lithos* 119: 412–26.
Merriam-Webster 1997. *Merriam Webster's Geographical Dictionary*. Third den. Springfield, MA: Merriam-Webster Inc.
Mikhaylov, V.A. 2009a. "Gidromorfnyy landshaftny urovyen'" ["The level of hydromophic landscapes"]. In *Sovremenyie landshaftty Kryma i sopredel'nykh akvatoriy*, ed. Ye.A. Pozachyenyuk, 200–26. Simferopol: Biznyes-Inform.
——— 2009b. Nyeblagopriyatnyye protsyesy [Adverse processes]. In *Sovremenyie landshaftty Kryma i sopredel'nykh akvatoriy*, ed. Ye.A. Pozachyenyuk, 243–7. Simferopol: Biznyes-Inform.
Milanovskiy, Ye.Ye. 1997. "The 'prehistory' and initial stages in the history of educational geological practical work by MGU in the Bakhchisaray region of the mountainous region of the Crimea". *Vyestnik – Moskovskogo Universityeta. Syeriya Geologiya* 3: 3–6.
Mishnyov, V.G. 2005. "Ekodinamika lyesov Kryma" ["Ecodynamics of the Crimean forests"]. *Geopolitika i Geodinamika Regionov* 1: 99–114.
Mokrzhyetskiy, S.A. 1914. "Fauna Kryma" ["The fauna of Crimea"]. In *Krym.Putyevodityel'*, ed. K.Yu. Bumbyer, L.S. Vagin, N.N. Klyepyenin and V.V. Sokolov, 73–102. Simferopol: Tipografiya Tavrichyeskoy Gubyernii Zyemstva.
Molchanov, Ye.F. and N.I. Rubtsov. 1986. *Nikitskiy Botanichyeskiy Sad: K 175-lyetyu osnovanya* [*The Nikitskiy Botanical Garden: Towards the 175th anniversary of its foundation*]. Kiev: Naukova Dumka.
Moon, D. 2013. *The Plough that Broke the Steppes: Agriculture and Environment on Russia's Grasslands*. Oxford: Oxford University Press.
Morozova, O.V. and Kozharinov, A.V. 2001. "Vliyaniye istorichyekikh faktorov na raspryedyelyeniye vidovogo bogatsva flory vostochnoy Yevropy" ["Influence of historic factors in the distribution of species richness in Eastern Europe"]. *Izvyestiya Akademii Nauk, Gyeograficbyeskaya Syeriya* 5: 39–50.
Morris, I. 2003. "Mediterraneanization". *Mediterranean Historical Review*, 18: 30–55.
Moysyeyev, L.A. 1926. "Slyedy irrigatsii, myelyoratsii i vodosnabzheniya dryevnyego Khyersonesa v Gyerakleiskom Poluostrovye" ["Vestiges of irrigation, land improvement and water supply in ancient Chersonesos in the Heraklean Peninsula"]. *Bulletin de la Societé de Nature de Crimée* 9: 115–22.
Muratov, M.V. 1974. *Krym I Geologicheskoye Opisaniye. Gyeologiya SSSR T. 1* [*Crimea I. Geological description. Geology of the USSR, vol.1*]. Moscow: Ministry of Geology of the USSR.
Muscheler, R. and Fischer, E. 2012. "Solar and volcanic forcing of decadal- to millennial-scale climatic variations". In *The SAGE Handbook of Environmental Change*, ed. J.A. Matthews, P.J. Bartlein, K.R. Briffa, A.G. Dawson, A. de Vernal, T. Denham, S.C. Fritz, and F. Oldfield, 444–70. Los Angeles: SAGE.
Myers, N.R.A., Mittermeier, C.G., da Fonseca, G.A.B. and Kent, J. 2000. "Biodiversity hot spots for conservation Priorities". *Nature* 403: 833–58.
Naveh, Z. 1982. "Mediterranean landscape evolution and degradation as multivariate biofunctions: theoretical and practical implications". *Landscape Planning* 9: 125–46.
Nazarov, N., Cook, H.F. and Woodgate, G. 2004. "Water pollution in Ukraine: The search for possible solutions". *Water Resources Development* 20: 205–18.
Nikolaenko, G.M. 2006. "The chora of Tauric Chersonesos and the cadastre of the 4th–2nd century BC". In *Surveying the Greek Chora: Black Sea Region in a Comparative Perspective*, ed. P. G. Bilde and V.F. Stolba, 151–74. Aarhus, Denmark: Aarhus University Press.
Novičenkova, N.G. 2008. "Mountainous Crimea: A frontier zone of ancient civilization". In *Meetings of Cultures: Between Conflicts and Coexistence*, ed. P.G. Bilde and J.H. Petersen, 287–302. Aarhus, Denmark: Aarhus University Press.
Novik, N.N. 1990. "O vozmozhnosti opryedyeleniya vosrasta ryechnykh tyerras po moshchnosti vyvyetryelogo sloya v gal'kakh alliyuviya" ["On the possibility of determining ages of river terraces through the weathering layer of clasts in alluvium"]. *Gyeologicheskiy Zhurnal* 5: 57–62.
Oğuz, T., Dippner, J.W. and Kaymaz, Z. 2006. "Climatic regulation of Black Sea hydrometeorological and ecological properties at interannual-to-decadal time scales". *Journal of Marine Systems* 60: 235–54.

Oğuz, T. and Öztürk, B. 2011. "Mechanisms impending natural mediterranization process of Black Sea Fauna". *Journal of Black Sea/Mediterranean Environment* 17: 234–53.

Olenkovsky, M.P. 2000. *Paleolit ta Mezolit Prysyvashshya. Problemy Epigravetu Ukrayiny* [*The Paleolithic and Mesolithic of the Sivash region, The issue of the Epigravettian of Ukraine*]. Kherson, Ukraine: Vydavnytstvo Prydniprov'ya (in Ukrainian).

Olifyerov, A.N. 2011. *Yayla: Gyeografiya, lyes, voda* [*The Yaila: Geography, forest, water*]. Simferopol: Biznyes-Inform.

Olifyerov, A.N. and Timchenko, Z.N. 2009. "Ozernye landshafty Kryma" ["Lacustrine landscapes of Crimea"]. In *Sovremenyie landshaftty Kryma i sopredel'nykh akvatoriy*, ed. Ye.A. Pozachyenyuk, 482–96. Simferopol: Biznyes-Inform.

Osadchaya, T.S., Osvianiy, E.I., Kemp, R., Romanov, A.S., and Ignatieva, O.G. 2003. "Organic carbon and oil hydrocarbons in bottom sediments of Sevastopol Bay (the Black Sea)". *Mor'skiy Ekologichny Zhurnal* 2: 94–101.

Ostapchuk, V. 2001. "The human landscape of the Ottoman Black Sea in the face of the Cossack naval raids". *Oriente moderno* 20: 23–95.

Pallas, P.S. 1812. *Travels through the Southern Provinces of the Russian Empire in the Years 1793 and 1794*. London: John Stockdale.

Pánek, T., Hradecký, J., Smolková, V. and Šilhán, K. 2008. "Gigantic low-gradient landslides in the northern periphery of the Crimean Mountains". *Geomorphology* 95: 449–73.

Parishkura, S.I. 1978. "Palyeogeograficheskiye usloviya formirovaniya pliotsyenovykh otlozhyeniy pryedgornogo Kryma po sporovo-pyltsevym dannym" ["Paleogeographical conditions of Pliocene deposits of piedmont Crimea using pollen and spores"]. *Fizicheskaya Geografiya i Geomorfologiya* 20: 79–86

Pasynkov, A.A. 2003a. "Tektonika" ["Tectonics"]. In *Atlas: Avtonomnaya Respubklika Krym* [*Atlas: The Autonomous Republic of Crimea*], ed. M.B. Bagrov and L.G. Rudenko, 19. Kiev and Simferopol: Tavrida National University, Crimean Center of the National Ukrainian Academy of Sciences, Institute of Geography of the NAN and Institute for Transfer Technology.

——— 2003b. "Chyetvyortichniye otlozheniya" ["Quaternary deposits"]. In *Atlas: Avtonomnaya Respubklika Krym* [*Atlas: The Autonomous Republic of Crimea*], ed. M.B. Bagrov and L.G. Rudenko, 20. Kiev and Simferopol: Tavrida National University, Crimean Center of the National Ukrainian Academy of Sciences, Institute of Geography of the NAN and Institute for Transfer Technology.

Patou-Mathis, M. 2005. "Analyses archéozoologiques des unités V et VI de Kabazi II". In *Kabazi II: Last Interglacial Occupation, Environment & Subsistence. The Palaeolithic Sites of Crimea. Volume 1*, ed. V. Chabai, J. Richter and T. Uthmeier, 77–98. Simferopol-Cologne: National Academy of Sciences of Ukraine, Crimean Filial and Institute of Prehistory and Protohistory, University of Cologne.

——— 2007. "Analyses archéozoologiques des unités III et IV de Kabazi V". In *Kabazi V: Interstratification of Micoquian and Levallois-Mousterian Camp Sites. The Palaeolithic Sites of Crimea, Volume 3, Part 1,* ed. V. Chabai, J. Richter and T. Uthmeier, 97–128. Simferopol-Cologne: Shlyakh.

Pavlov, S.S., Roerink, G.J., Hellegers, P.J. and Popovych, V.F. 2006. "Irrigation performance assessment in Crimea, Ukraine". *Water Resources Development* 22: 61–78.

Permyakov, V.V. and Maidanovich, J.A. 1984. "The geological structure of the fold regions of the Ukraine (general overview)". In *International Geological Congress 27th Session Moscow, Field Trip Guidebook*, 3–32. Kiev: Naukova Dumka.

Petit, R.J., Brewer, S., Bordács, S., Burg, K., Cheddadi, R., Coart, E., Cottrell, J. et al. 2002. "Identification of refugia and post-glacial colonisation routes of European white oaks based on chloroplast DNA and fossil pollen evidence". *Forest Ecology and Management* 156 (1–3): 49–74.

Pettit, P.B. 1998. "Middle and Early Upper Paleolithic Crimea: the radiocarbon chronology". In *Préhistoire d'Anatolie: Genèse de deux mondes*, ed. M. Otte, 329–38. Liège: ERAUL.

Pevzner, M.A., Syemyenyenko [= Semenenko], V.N., Vangeigeim, E.A., Sadchikova, T.A., Kovalyenko, V.A. and Lyul'eva, S.A. 2004. "O morskom genezise i ponticheskom vozrastye otlozheniy opornogo razreza Lyubimovka v Krymu" ["On the marine origin and Pontic age of the Ljubimovka deposits in Crimea"]. *Stratigrafiya i gyeologicheskaya korryelyatsiya* 12 (5), 96–106.

Podgorodyetskiy, P.D. 1988. *Krym: Priroda [Nature of Crimea]*. Simferopol: Tavriya.

Pozachyenyuk, Ye. A. (ed.). 2009a. *Sovremenyie landshaftty Kryma i sopredel'nykh akvatoriy [Modern landscapes of Crimea and surrounding seas]*. Simferopol: Biznyes-Inform.

——— 2009b. "Ekotsyentricheski-setevaya" ["Ecological net"]. In *Sovremenyie landshaftty Kryma i sopredel'nykh akvatoriy*, ed. Ye.A. Pozachyenyuk, 185–95. Simferopol: Biznyes-Inform.

——— 2012. "Obshchaya kharakteristika rastityelnost'" ["General characteristics of vegetation"]. Unpublished assessment report on the Karan' Plateau provided by the author.

Pozachyenyuk, Ye.A. and Grishankov, G.E. 1998. "Shiblyakopodovnyye soobshchestvo Kryma" ["Communities similar to the shiblyak"]. In *Biologicheskoye i Landshaftnoye raznoobraziye Kryma: Problemy i perspektivy*, ed. L.G. Apostolov, V.A. Bokov and A.I. Dulitskiy, 80–3. Simferopol: Tavriya.

Prat, S., Péan, S.C., Crépin, L., Drucker, D.G., Puaud, S.J., Valladas, H., Lázničková-Galetová, M., Van der Plicht, J. and Yanevich, A. 2011. "The oldest anatomically modern humans from far southeastern Europe: direct dating, culture and behavior". *PLoS ONE* 6: 1–13.

Prokudin, Yu.N., Vovk, A.G., Pyetrova, O.A., Yermolyenko, Ye.D. and Vernichyenko, Yu.V. 1977. *Zlaki Ukrainy [The grasses of Ukraine]*. Kiev: Naukova Dumka.

Provan, J. and Bennett, K.D. 2008. "Phylogeographic insight into cryptic glacial refugia". *Trends in Ecology and Evolution* 23: 564–71.

Prusakov, A.A., and Kozlov, Ye. D. 1999. *Kniga Ryekordov Kryma* [Crimea book of records]. Simferopol: Sonat.

Radchenko, V.N. and Aleyev, M.Y. 2000. "Environmental and social impacts of management approaches in Sevastopol Bay in a historic retrospective: a case study from the Black Sea". *Ocean and Coastal management* 43: 793–817.

Raven, P.H. 1971. "The relationships between 'Mediterranean' Floras". In *Plant Life of South-West Asia*, ed. P.H. Davies, P.C. Harper and I.C. Hedge, 213–24. Edinburgh: The Botanical Society of Edinburgh.

Richter, J. 2005. "Hasty foragers: The Crimea Island and Europe during the Last Interglacial". In *Kabazi II : Last Interglacial Occupation, Environment and Subsistence. The Palaeolithic Sites of Crimea, Volume 1*, ed. V. Chabai, J. Richter and T. Uthmeier. 275–86, Simferopol-Cologne: National Academy of Sciences of Ukraine, Crimean Filial and Institute of Prehistory and Protohistory, University of Cologne.

Rikli, M.A. 1943. *Das Pflanzenkleid der Mittelmeerländer* (3 vols) Bern: H. Huber.

Roberts, N. 2012. "Climate change and the Justinianic Plague: an intercomparison of high-resolution lake sediment and documentary records (Abstract of the XVIII INQUA Congress, 21st-27th July, 2011, Bern, Switzerland)". *Quaternary International* 279–80: 407.

Roberts, N., Moreno, A., Valero Garcés, B.L., Corella, J.P., Jones, M., Allcock, S., Woodbridge, J., Morellón, M., Luterbacher, J., Xoplaki, E. and Türkeş, M. 2012. "Paleolimnological evidence for an east-west climate see-saw in the Mediterranean since AD 900". *Global and Planetary Change* 84–5: 23–4.

Robinson, J. 2006. *The Oxford Companion to Wine*. New York: Oxford University Press.

Rosen, W. 2008. *Justinian's flea: the First Great Plague and the End of the Roman Empire*. New York: Penguin Books.

Rubtsov, N.I. 1972. *Opryedyelityel' vyschikh rastyeniy Kryma [Identification manual to the vascular plants of Crimea]*. Leningrad: Izdatyel'stvo Nauka.

——— 1978. *Rastityel'ny mir Kryma [The vegetation world of Crimea]*. Simferopol: Tavriya.

——— 1980. "Geograficheskiye svyazi flory Kryma i gipotyeza Pontidy" ["Geographic links of Crimean flora and the Pontida hypothesis"]. *Priroda* 1980 (1): 50–8.

Rukhlov, N.V. 1915. *Obzor ryechnykh dolin gornoy chasti Kryma [Survey of the mountain rivers of Crimea]* Pyetrograd (St Petersburg): Tipografiya V.O. Kirshbauma.

Ryan, W.B.F., Major, C.O., Lericolais, G. and Goldstein, S.L. 2003. "Catastrophic flooding of the Black Sea". *Annual Review Earth and Planetary Sciences* 31: 525–54.

Ryan, W.B.F. and Pitman III, W.C. 1998. *Noah's Flood: The New Scientific Discoveries about the Event that Changed History*. New York: Simon & Schuster.

Rybintsev, V. 1995. "Viticulture and oenology in the Ukraine: A survey of their history, ecology and economics". *Journal of Wine Research* 6: 35–47.

Rybintsev, V., and Petkilyova, G. 1992. "Trends in the development of viticulture and oenology in the Crimea". *Journal of Wine Research* 3: 137–42.

Saenko, O.A., Knysh, V.V. and Korotaev, G.K. 2000. "Reconstruction of the seasonal climate of the Black Sea on the basis of available hydrological data". *Physical Oceanography* 11: 23–46.

Saprykin, S.J. 1994. *Ancient Farms and Land Plots on the Khora of Khersonesos Taurike*. Amsterdam: J.C. Gieben.

Savelya, O.Y. 1996. "Arkheologicheskiye materially k istorii Gerakleiskogo Poluostrova dokolonizatsionnogo perioda" ["Archaeological materials for the history of the Heraklean Peninsula before the Colonization Period"]. *Khyersonyeskiy Sbornik* 7: 13–18.

Šceglov, A.N. 2002. "Introduction". In *Panskoye I, Vol. 1, U6 the Monumental Building*, ed. L. Hannestad, V.F. Stolba, and A.N. Šceglov, 11–25. Aarhus, Denmark: Aarhus University Press.

Schreg, R. 2010. "Landschaftsärchaologische Forschungen zu den Höhensiedlungen der südwestlichen Krim und ihrem Umland". In *Byzanz: Das Römerreich im Mittelalter. Part 3 Peripherie und Nachbarshaft*, Monographien des Römisch-Germanisches Zentralmuseums 84 (3), ed. F. Daim, and J. Drauschke, 95–109. Mainz: Römsich-Germanisches Zentralmuseum.

——— 2011. "Ubernutzung einer Berglandschaft? – Umweltarchäologie auf der südwestlichen Krim". Final research report to the Fritz Thyssen Foundation. Unpublished. Mainz, Germany.

Schulz, H.M., Vakarcs, G. and Magyar, I. 2005. "The birth of the Paratethys during the Early Oligocene: From Tethys to an ancient Black Sea analogue?" *Global and Planetary Change* 49: 163–76.

Seymour, H.D. 1855. *Russia on the Black Sea and Sea of Azof: A Narrative of Travels in the Crimea and Bordering Provinces*. London: John Murray.

Shamko, E. 1963. "Iz istorii vinogradstva i sadovodstva v Krymu" ["From the history of winemaking and horticulture in Crimea"]. *Vinogradstvo i Sadovodstvo Kryma* 1: 41–3.

Sharygin, S.A. 1995. "The batrachofauna of the southern shore of Crimea". In *Amphibian Populations in the Community of Independent States: Current Status and Decline*, ed. Kuzmin, S.L., Dodd Jr., C.K. and M.M. Pikulik, 45–8. Moscow: Informsvyazizdat.

Shchyekina, N.A. 1979. *Istoriya flory I rastityel'nosti yuga Yevropyeyskoy chasti SSSR v pozdnyem Miotsyenye-rannyem Pliotsyenye* [*Floristic and vegetational history of the southern European part of the USSR during the late Miocene and early Pliocene*]. Kiev: Naukova Dumka.

Scherbak, N.N., 1966. *Zyemnovodnyye i pyeresmykayushchyesya Kryma* [*Amphibians and reptiles of Crimea*]. Kiev: Naukova Dumka.

Shchyerbakov, F.A., Kuprin, P.N., Zabelina, Ye.K. and Parunin, O.B. 1977. "Paleogeography of the Azov and Black seas in the Late Pleistocene and Holocene". In *Paleogeography and Pleistocene Deposits of the Southern Seas of the USSR*, ed. P.A. Kalinin and F.A. Shchyerbakov. (Pages missing.) Moscow: Izdatel'stvo Nauka.

Shchyerbakov, F.A., Kuprin, P.N., Polyakov, A.S., Balandin, Yu. G., Ivanov, G.I. and Rotar', M.F. 1976. "Shyel'f syevyero-zapadnoy chasti Chyornogo Morya v pozdnyem Plyeystotsyenye-Golotsyenye" ["Shelf of the northwestern part of the Black Sea during the Late Pleistocene and Holocene"]. *Chyetvyortichny Pyeriod* 16: 141–52.

Shelyag-Sosonko, Yu. R., Osychnyuk, V.V. and Andriyenko, T.L. 1982. *Gyeografiya rastitityel'nogo pokrova Ukrainy* [*Geography of the vegetation cover of Ukraine*]. Kiev: Naukova Dumka.

Shilik, K.K., 1997. "Oscillations of the Black Sea and ancient landscapes". In *Landscapes in Flux: Central and Eastern Europe in Antiquity*, Colloquia Pontica, vol. 3, ed. J. Chapman and P. Dolukhanov, 115–29. Oxford: Oxbow Books.

Shostakovich, V.B. 1934. "Iloviye otlozheniya i pyeriodicheskiye kolyebaniya v yavleniyakh prirody" {"Silt lake deposits and periodical oscillations of natural phenomena"}. *Zapiski Gosudarstvyenogo Gidrologicheskogo Instituta* 13: 95–140.

Shuisky, Y. 2007. "Climate dynamics, sea-level change and shoreline migration in the Ukrainian sector of the Circum-Pontic Region". In *The Black Sea Flood Question: Changes in Coastline, Climate and Human Settlement*, ed. V. Yanko-Hombach, A.S. Gilbert, N. Panin and P.M. Dolukhanov, 251–78. Berlin and New York: Springer.

Shults, P.N. 1967. "Antropomorfnaya styela sarmatskogo kruga, naydydennaya na Arabatskoy Strel'kye" {"Anthropomorphic stela of Sarmatian age found in the Arabastkaya Stryel'ka"}. *Zapiski Odesskogo Arkhyeologicheskogo Obshchyestva* 2: 196–201.

Shumskiy, V.M. 2009. "Khozyaystvyeny komplyeks" {"The economic complex"}. In *Sovremenyie landshaftty Kryma i sopredel'nykh akvatoriy*, ed. Ye.A. Pozachyenyuk, 141–63. Simferopol: Biznyes-Inform.

Siokhin, V. and Kostyushyn, V. 2000. *Sivash: The Lagoon Between Two Seas*. Kiev: Wetlands International.

Slavin, V.I. 1975. *Sovreemennyye geologicheskoye protsyessy v yugo-zapadnom Krymu* {*Modern geological processes in southwestern Crimea*}. Moscow: Izdatyel'stvo Moskoskogo Universityeta.

Smekalova, T.N. and Smekalov, S.N. 2006. "Ancient roads and land division in the chorai of the European bosporos and chersonesos on the evidence of air photographs, mapping and surface survey". In *Surveying the Greek Chora. The Black Sea Region in a Comparative Perspective*, ed. P.G. Bilde and V.F. Stolba, 207–48. Aarhus, Denmark: Aarhus University Press.

Solomina, O., Davi, N., D'Arrigo, R. and Jacoby, G. 2005. "Tree-ring reconstruction of Crimean drought and lake chronology correction". *Geophysical Research Letters* 32: L19704.

Sorochan, S.B., Zubar, V.M. and Marchenko, L.V. 2000. *Zhizn' i gibyel' Khersonesa* {*Life and decline of Chersonesos*}. Kharkov, Ukraine: Maydan.

Spencer, E. 1855. *Turkey, Russia, the Black Sea and Circassia*. London: George Routledge.

Stanko, V.N. 2007. "Fluctuations in the level of the Black Sea and Mesolithic settlement of the northern Pontic area". In *The Black Sea Flood Question: Changes in Coastline, Climate and Human Settlement*, ed. V. Yanko-Hombach, A.S. Gilbert, N. Panin and P.M. Dolukhanov, 371–85. Berlin and New York: Springer.

Stashchuk, M.F., Suprichev, V.A. and Khitraya, T. 1964. *Mineralogiya, gyeokhimiya i usloviya formirovaniya donnykh otlozhyeniy Sivasha* {*Mineralogy and geochemical conditions and the formation of the Sivash sediments*}. Kiev: Naukova Dumka.

Stepanchuk, V.S., Ryzhov, S., Rekovets, L. and Matviishina, Zh. 2010. "The Lower Paleolithic of Ukraine: current evidence". *Quaternary International* 223–4: 131–42.

Steppe Bulletin Editorial. 2013. "Novosti: Pryecyedyenty Nakazanija za nyezakonnoye oblyesyeniye styepyej v Ukrainye" {"News: Precedents of punishment on illegal afforestation in the steppes of Ukraine"}. *Styepnoy Byulyetyen'* 38: 32–3.

Strzhyelyetskiy, S.F. 1961. *Klyery Khyersonyesa Tavrichyeskogo. K istorii dryevneyvo zyemlyedyeliya v Krymu* {*Farms of Tauric Chersonesos. Towards a history of ancient agriculture in Crimea*}. *Khyersonyeskiy Sbornik* 6. Simferopol: Krymizdat.

Subetto, D.O., Gerasimenko, N.P., Bakhmutov, V.G., Neustreva, I.Yu, Sapelko, T.V., Kuznetsov, D.D., Davidova, N.N., Ludikova, A.V. and Stolba, V.F. 2009. "Rekonstruktsiya paleografichnykh umov zakhidnogo Krymu u piznomu golotseni za litologichnymy i paleontologichnymy materialamy vivchennya ozer" {"Reconstruction of paleogeographic conditions of Western Crimea in the late Holocene through the lithological and paleontological study of lakes"}. *Fizychna Geografiya ta Geomorfologiya* 56: 299–310 (in Ukrainian).

Sumarokov, P.I. 1800. *Putyeshyestviye po vsyemu Krymu i Bessarabii v 1799 godu* {*Travel through all Crimea and Bessarabia in the year 1799*}. Moscow: Univyersityetskaya Tipografiya.

——— 1803–5. *Dosugi krymskogo sud'i ili vtoroye putyeshestviye v Krymu* {*Leisure of a Crimean judge*}. St Petersburg: Imperatorskaya Tipografiya.

Tamaychuk, A.N., Andreyeva, O.A. and Penno, M.B. 2009. "Landshafty Chernogo Morya" ["Landscapes of the Black Sea"]. In *Sovremenyie landshaftty Kryma i sopredel'nykh akvatoriy*, ed. Ye.A. Pozachyenyuk, 497–549. Simferopol: Biznyes-Inform.

Tolstoy, L. 1986. *The Sebastopol Sketches* (translation by David McDuff). London: Penguin Books.

Tomaselli, R. 1977. "The degradation of the Mediterranean maquis". *Ambio* 6: 356–62.

Trelogan, J. 2001. "Mapping the ancient territory". In *The Study of Ancient Terriroties: 2001 Annual Report of the Institute of Classical Archaeology*, ed. J.C. Carter, 33–8. Austin, TX: Institute of Classical Archaeology, University of Texas at Austin.

Tringham, R. 1969. "Animal domestication in the Neolithic cultures of the south-west part of European USSR". In *The Domestication and Exploitation of Plants and Animals*, ed. P.J. Ucko, and G.W. Dimbleby, 381–92. Chicago: Aldine Publishing Company.

Turnock, D. 2001. *East Central Europe and the Former Soviet Union: Environment and Society.* London: Arnold.

Twain, M. 1966. *The Innocents Abroad.* New York: Putnam Classic.

Uthmeier, T., Ickler, S. and Kurbjuhn, M. 2008. "Site catchment analysis in the Late Crimean Middle Palaeolithic: a GIS based approach". In *Kabazi V: Interstratification of Micoquian & Levallois-Mousterian Camp Sites. The Palaeolithic Sites of Crimea. Volume 3, Part 2*, ed. V.P. Chabai, J. Richter and T. Uthmeier, 481–508. Simferopol-Cologne: Shlyakh.

Vakhrushyev, B.A. and Klyukin, A.A. 2001. Kriogyennye protsyessy krymskikh Yayl [Cryogenic processes of the in the Crimean yailas. *Gyeomorfologiya* 2001–2: 48–54.

Vakhrushyeva, L.P. 2009. "Ekosistyemy" ["Ecosystems"]. In *Sovremenyie landshaftty Kryma i sopredel'nykh akvatoriy*, ed. Ye.A. Pozachyenyuk, 128–41. Simferopol: Biznyes-Inform.

Vakhrushyeva, L.P. and Yena, A.V. 2003. "Species in the Ukrainian Red Book". In *Atlas: Avtonomnaya Respubklika Krym* [*Atlas: The Autonomous Republic of Crimea*], ed. M.B. Bagrov and L.G. Rudenko, 34–5. Kiev and Simferopol: Tavrida National University, Crimean Center of the National Ukrainian Academy of Sciences, Institute of Geography of the NAN, and Institute for Transfer Technology.

Vasiliev, A.A. 1936. *The Goths in the Crimea.* Cambridge, MA: The Mediaeval Academy of America.

Vasilyuk, A. 2009. "Sokhranit' ili oblyesit'?" ["Preserve or plant trees?"] *Styepnoy Byulyetyen'* 26: 38–40.

Veinstein, G. 1986. "From the Italians to the Ottomans: The case of the northern Black Sea coast in the sixteenth century". *Mediterranean Historical Review*, 1: 221–37.

Vita-Finzi, C. 1969. *The Mediterranean Valleys: Geological Changes in Historical Times.* London: Cambridge University Press.

Vodop'yanova, T.D. 1986. "Dikorastuyushchiye vyechnozelyonye sryediszemnomorskiye rastyeniya na syevyernoy granitsye svoyevo raspostraneyeniya" ["Evergreen Mediterranean plants on their northernmost distribution"]. *Botanichyeskiy Zhurnal* 7: 222–31.

Vogiatzakis, I.N. 2012. *Mediterranean Mountain Environments.* Chichester: Wiley.

Vremir, M.M. 2004. "New approach on the saiga antelopes (Saiga tatarica/borealis; Bovidae) from the Late Pleistocene and Holocene in Crimea, Ukraine". *Studia Universitatis Babeş-Bolyai, Geologia* 49 (2): 105–8.

Vremir, M.M. and Ridush, B. 2002. "Recent paleontological investigations in some caves of the Crimean mountain range (SE 2Ukraine)". *Theoretical and Applied Karstology* 15: 125–32.

Vul'f, Ye.V. 1925. *Rastityelnost' vostochnikh yayl Kryma* [*Vegetation of the eastern yailas of Crimea*]. Moscow: Novaya Dyeryevnya.

——— 1926. "Proiskhozhdyeniya flory Kryma" ["Origin of the Crimean flora"]. *Zapiski Obshchyestva Yestyestoispytatyelyey i Lyubityelyey Prirody* 9: 81–108.

——— 1944. "Istorichyeskaya gyeografiya rastyeniy. Istoriya flor zyemnogo shara" ["Historical geography of plants. History of the flora of the world"]. Moscow: Izdatyel'stvo Akadyemii Nauk SSSR.

Vul'f, Ye.V., Kalayda, F.K. and Plotnitskiy, G.A. 1916. *Kultura masliny Olea europaea L. na Yuzhnom Byeryegu Kryma* [*The cultivation of olive Olea europaea L. in the south coast of Crimea*]. Yalta: Imperial Nikitsky Botanical Gardens.

Vyeklich, M.F. 1990. *Stratigraficheskaya skhyema chyetvyortichnykh otlozhyeniy Ukrainy* [*Stratigraphic scheme of the Quaternary deposits of Ukraine*]. Kiev: Gosgeolkom Ukrainy.

Vyeklich, M.F. and Siryenko, N.A. 1976. *Pliotsyen i Plyeystotsyen lyevobeyeryezhya nizhnyego Dnyepra i ravnonnogo Kryma* [*The Pliocene and Pleistocene of the left bank of the Dnieper and the plains of Crimea*]. Kiev: Naukova Dumka.

Vyed', I.P. 1983. "Klimatopy rastitel'nikh obshchestv gornogo Kryma" ["Climatic patterns of the vegetation communities of the Crimean Mountains"]. *Izvyestiya Akademii Nauk SSSR, Seriya Geograficheskaya* 3: 83–9.

——— 2000. *Klimatichyeskiy Atlas Kryma* [*Climate atlas of Crimea*]. Simferopol: Tavriya-Plyus.

Vyeklich, M.F., Siryenko, N.A., Maviishina, Zh.N., Mel'nichuk, I.V. Gerasimenko, N.P., Pyeredyeriy, V.I. and Turlo, S.I. 1993. *Stratigrafichyeskaya skhyema chyetvyortichnykh otloshyeniy Ukrainy* [*Stratigraphic scheme of the Quaternary Deposits of Ukraine*]. Kiev: Naukova Dumka.

Vyeklich, M.F., Sirenko, N.A. and Matviishina, Zh.N. 1993. *Stratigraficheskaya skhyema plyestotsyenovykh otlozhyeniy Ukrainy* [*Stratigraphic scheme of the Pleistocene deposits of Ukraine*]. Kiev: Goskomityet geologii Ukrainy.

Walter, H. 1943. *Die Krim. Klima, Vegetation und landwirtschaftliche Erschliessung*. Berlin: Verlag C.V. Engelhard.

——— 1974. *Die Vegetation Osteuropas, Nord- und Zentralasiens*. Stuttgart: Fischer.

——— 1980. *Bekenntnisse eines Ökologen: Erlebtes in acht Jahrzehnten und auf Forschungsreisen in alles Erdteilen*. Stuttgart: Fischer.

Wendland, F. 1992. *Peter Simon Pallas (1741–1811)*. Materielen einer Biographie. Berlin: Walter de Gruyter.

Whelis, M. 2002. "Biological warfare at the 1346 siege of Caffa". *Emerging Infectious Diseases* 8: 971–5.

Winslow, J.C., Hunt Jr., E.R. and Piper, S.C. 2003. "The influence of seasonal water availability on global C3 versus C4 grassland biomass and its implications for climate change research". *Ecological Modelling*: 153–73.

Woodbridge, J. and Roberts, N. 2011. "Late Holocene climate of the eastern Mediterranean inferred from diatom analysis of annually-laminated lake sediments". *Quaternary Science Reviews* 30: 3381–92.

Wulff [= Vul'f], E.V. 1932. "The beech in the Crimea. Its systematic position and origin". In *Die Buchenwälder Europa*, Veröffentlichungen des botanischen Instituts 8, ed. E. Rübe, 223–60. Zurich: Stiftung Rübel.

Yakobson, A.L. 1970. *Rannyesryednyevyekovye syel'skiye posyelyeniya yugo-zapadnoy Tavriki* [*Early rural settlements of southwestern Crimea*]. Leningrad: Nauka.

Yakovyenko, I.M. 2009. "Khosyaystvyenaya spyesyalizatsiya" ["Economic specialization"]. In *Sovremenyie landshaftty Kryma i sopredel'nykh akvatoriy*, ed. Ye.A. Pozachyenyuk, 279–92. Simferopol: Biznyes-Inform.

Yakovlyev, A.N. 2009. "Khozyaystvyenaya spyetsialozatsiya" ["Economic specialization"]. In *Sovremenyie landshaftty Kryma i sopredel'nykh akvatoriy*, ed. Ye.A. Pozachyenyuk, 313–16. Simferopol: Biznyes-Inform.

Yanata, A.A. 1913. "Ochyerk rastityel'nosti Kryma" ["Essay on the vegetation of Crimea"]. In *Krym. Putyevodityel'*, ed. K.Yu. Bumbyer, Vagin, L.S., Klyepyenin, N.N. and Sokolov, V.V, 41–72. Simferopol: Tipografiya Tavrichyeskoy Gubyernii Zyemstva.

Yanko-Hombach, V.V. 2007. "Controversy over Noah's Flood in the Black Sea: geological and foraminiferal evidence from the shelf". In *The Black Sea Flood Question: Changes in Coastline, Climate and Human Settlement*, ed. V. Yanko-Hombach, A.S. Gilbert, N. Panin, and P.M. Dolukhanov, 149–205. Berlin and New York: Springer.

Yanko-Hombach, V., Gilbert, A.S., Panin, N. and Dolukhanov, P.M. (eds). 2007. *The Black Sea Flood Question: Changes in Coastline, Climate and Human Settlement*. Berlin and New York: Springer.

Yanuchevitch, Z.V., Nikolaenko, G. and Kuzminova, N. 1985. "La viticulture à Chersonèse de Taurique aux IVe-IIe siècles av. n.è. d'après les recherches archéologiques et paléoethnobotaniques". *Revue d'Archéologie* 1: 115–22.

Yavuz, V., Akçar, N. and Schlüchter, C. 2007. "The frozen Bosphorus and its paleoclimatic implications based on a summary of historical data". In *The Black Sea Flood Question: Changes in Coastline, Climate and Human Settlement*, ed. V. Yanko-Hombach, A.S. Gilbert, N. Panin and P.M. Dolukhanov, 633–49. Berlin and New York: Springer.

Yena, A.V. 1978. "Prirodo-tyerritorial'nyye komplyeksov Subsryedizyemnomor'ya. Problyemy okhrany ikh prirody i napravlyeniya ratsional'nogo ispol'zovanii" {"Natural territorial complexes of the Sub-Mediterranean region. Issues concerning its nature and directions of rational use"}. Dissertation for the degree of candidate to doctor of geographical sciences. Thesis, Ukrainian Academy of Sciences of the USSR, Institute of Geophysics, Geography Section. Kiev.

——— 2007. "Floristic endemism in the Crimea". *Fritschiana* 55: 1–8.

——— 2012. *Prirodnaya Flora Krymskogo Poluostrova* {*Spontaneous flora of the Crimean Peninsula*}. Simferopol: Orianda.

Yena, A.V. and Yena, A.V. 2008. *Lotsiya Kryma* {*The pilot's book of Crimea*}. Simferopol: Biznyes Inform.

——— 2010. *Kuesty Krymskogo Pryedgoriya* {*Cuestas of the Crimean Piedmont*}. Simferopol: Orianda.

Yena, V.G., Yena, A.V. and Yena, A.V. 2004a. "Katastrofichyeskiye poslyedstviya poslyednyey transgressii dlya landshaftnogo I floristichyekogo raznoobraziya Krymskogo subsryedizyemnomor'ya" {"Catastrophic consequences of the last transgression on the landscape and floristic diversity of the Crimean Sub-Mediterranean region"}. In *Ukrayina: heohrafichni problem stalogo rozvytku*, ed. P.G. Shchishchenko, 229–31. Kiev: VGL Obriyi (chapter in Russian, publication in Ukrainian).

——— 2004b. *Zapovyednyye Landshafty Kryma* {*Protected landscapes of Crimea*}. Simferopol: Biznyes-Inform.

——— 2005. "'Stankiewicz Pine' in Crimea: Some new taxonomical, chorological and paleolandscape considerations". *Dendrobiology* 53, 63–9.

——— 2007. *Otkryvatyeli zyemli Krymskoy* {*The discoverers of the Crimean land. Essays of the 2,500-years of nature research in Tauria*}. Simferopol: Biznyes-Inform.

Yergina, Ye.I. 2009. "Klimat" {"Climate"}. In *Sovremenyie landshaftty Kryma i sopredel'nykh akvatoriy*, ed. Ye.A. Pozachyenyuk, 63–78. Simferopol: Biznyes-Inform.

Zakharov, S.A. 1929. *Kurs pochvovyedenija. S prilozhyeniyem klassifikacionnykh tablits i skhematiheskoj karty pochvyennych zon SSSR, Kavkaza i SSHA*. {*Soil science course. With attched classification tables of the soils map of the USSR, the Caucasus and the USA*}. Moscow: Sel'khozgiz.

Zgurovskaya, L. 1996. *Krym rasskazy o rastyeniyakh i zhivotnykh* {*Crimea: tales about plants and animals*}. Simferopol: Biznyes-Inform.

Zohary, M. 1971. "The Phytogeographical Foundations of the Middle East". In *Plant Life of South-West Asia*, ed. P.H. Davies, P.C. Harper and I.C. Hedge, 41–52. Edinburgh : The Botanical Society of Edinburgh.

Zolotaryev, M.I. 1979. "New data on ancient navigation routes in the Pontus Euxinus". In *Problems of Greek Colonization in the Northern and Western Part of the Black Sea*, ed. O. Lordkipanidze, 94–1-100. Tbilisi: Metsinereba.

Zonn, I.S., Fashchuk, D.Y. and Ryabinin, A.I. 2008. "Environmental issues of the Black Sea". In *The Black Sea Environment. Handbook of Environmental Chemistry, Part 5*, ed. A.N. Kosarev, 407–21. Berlin: Springer-Verlag.

Zyenkovich, V.P. 1958. *Byerega Chyornogo i Azovskogo Moryey* {*The coasts of the Black and Azov Seas*}. Moscow: Gosudarstvyennoye Izdatyel'stvo Geograficheskoy Literatury.

References to Classical Authors

Herodotus (Herodotus and David Grene). 1987. *The Histories*. Chicago: University of Chicago Press. Nine books in one volume.

Pliny the Elder and Rackham, H. 1938. *Natural History*. Cambridge, Mass: Harvard University Press. 37 books in 10 volumes.

Strabo, Jones, H.L. and Sitlington Sterrett, J.R. 1917. *The Geography of Strabo*. London: W. Heinemann. 17 books in 8 volumes.

Online sources

Bondaryeva, L. 2010. Ugroza bioraznoobraziyu Sevastopol'skogo regiona – Zakaznik "Karan'skiy" nye sozdan! [Attack on the biodiversity of the Sevastopol region – the Karan' Zakaznik is not created!] *Natsionalnam Parkam – Buty!* Article available at: http://pryroda.in.ua/zapzf/ugroza-bioraznoobraziyu-sevastopolskogo-regiona-zakaznik-karanskij-ne-sozdan/ (last accessed: 14 August 2013).

Byeryegovaya, T. 2011. Gdye mozhno bezopasno kupat' v Krymu? [Where can one swim safely in Crimea?] *Syegodnya*, 11 June 2011. Available at: http://www.segodnya.ua/regions/krym/hde-mozhno-bezopacno-kupatcja-v-krymu.html (last accessed: 5 December 2013).

Dikorastuyushchiye Rastyeniya Kryma (Wild plants of Crimea) 2012, Beryoza povislaya. Available at: http://www.flora.crimea.ua/beresa/Betula.html (last accessed: 30 July 2013).

Flora Europaea Database 2012. Royal Botanical Garden, Edinburgh. Available at: http://rbg-web2.rbge.org.uk/FE/fe.html (last accessed: 30 July 2013).

Freshwater Ecoregions of the World 2013. *426 Crimean Peninsula*. The Nature Conservancy. Available at: http://www.feow.org/ecoregion_details.php?eco=426 (last accessed: 17 August 2013).

International Commission on Stratigraphy, 2013. *International Chronostratigraphic Chart v. 2013*.Available at: http://www.stratigraphy.org/ (last accessed: 30 June 2013).

International Panel on Climate Change 2007. *The Physical Science Basis: Contribution of Working Group I to the Fourth Assessment Report of the Intergovernmental Panel on Climate Change*. Available at: http://www.ipcc.ch/publications_and_data/ar4/wg1/en/contents.html (last accessed: 13 December 2013).

Litvinyenko, M. 2010. Zakaznik "Karan'skiy" – trudnaya sud'ba [The Karan' Zakaznik: a difficult existence]. Sevnews, available at: http://sevnews.info/rus/view-news/balaklava-Zakaznik-Karanskij-trudnaya-sudba/1860 (last accessed: 11 August 2013).

Parnikoza, I.Yu. and Vasilyuk, O.V. 2010. Ukrainian steppes: current state and prospects for conservation. Available at: http://pryroda.in.ua/step/doslidzhennya-rozrobki-sposterezhennya/ukrainian-steppes-current-state-and-perspectives-for-safe/ (last accessed: 7 December 2013).

Secretariat of the President of Ukraine 2009; Ukaz Prezident Ukrayiny pro stvorennya natsional'noho prirodnoho parku "Charivna Havan'" [Decree of the President of Ukraine on the creation of the National natural park "Magic harbor"] List of 11.12.2009 No 1037/2009. (In Ukrainian.) Available at: http://zakon4.rada.gov.ua/laws/show/1037/2009 (last accessed: 11 August 2013).

Slava Syevastopolya 2006. V Syevastopolye budyet zakaznik "Karan'skiy" [In Sevastopol there will be the Karan'skiy zakaznik]. *Slava Syevastopolyia*, 24 October 2006. Available at: http://www.slava.sebastopol.ua/2006.10.24/view/13774_v-sevastopole-budet-botanicheskiy-zakaznik-karanskiy.html (last accessed: 14 August 2013).

Zakirov, Z. 2013. Kitaytsy sobirayut postroit' v Krimu grandiozny port [The Chinese plan to construct a gigantic port in Crimea]. *Vesti Krym*, 6 December 2012. Available at: http://vesti.ua/krym/28469-kitajcy-v-krymu-hotjat-stroit-port (last accessed: 14 December 2013).

Zberezhemo Ukrayins'ki Stepy! 2010. *Ukrayins'ki Stepy* (Ukrainian steppes). Available at: http://pryroda.in.ua/step/ (last accessed: 7 December 2013).

INDEX

www.ingramcontent.com/pod-product-compliance
Ingram Content Group UK Ltd.
Pitfield, Milton Keynes, MK11 3LW, UK
UKHW020805290926
15470UKWH00003B/963

9 781784 530013